Digital Dragon

●●●

A volume in the series
CORNELL STUDIES IN POLITICAL ECONOMY
edited by Peter J. Katzenstein

A full list of titles in the series
appears at the end of the book.

Digital Dragon

...

High-Technology Enterprises in China

Adam Segal

A COUNCIL ON

FOREIGN RELATIONS BOOK

Cornell University Press

ITHACA AND LONDON

Copyright © 2003 by Cornell University

All rights reserved. Except for brief quotations in a review, this book, or parts thereof, must not be reproduced in any form without permission in writing from the publisher. For information, address Cornell University Press, Sage House, 512 East State Street, Ithaca, New York 14850.

First published 2003 by Cornell University Press

Printed in the United States of America

Library of Congress Cataloging-in-Publication Data

Segal, Adam, 1968–
    Digital dragon : high-technology enterprises in China / Adam Segal.
        p. cm. — (Cornell studies in political economy)
Includes bibliographical references and index.
    ISBN 0-8014-3985-X (cloth : alk. paper)
    1. High technology industries—China.   I. Title: High-technology Enterprises in China. II. Title. III. Series.
    HC79.H53 S436 2002
    338.4'76—dc21

                            2002007134

Cornell University Press strives to use environmentally responsible suppliers and materials to the fullest extent possible in the publishing of its books. Such materials include vegetable-based, low-VOC inks and acid-free papers that are recycled, totally chlorine-free, or partly composed of nonwood fibers. For further information, visit our website at www.cornellpress.cornell.edu.

Cloth printing 10 9 8 7 6 5 4 3 2 1

For Rona

Contents

Contents

Figures and Tables

Acknowledgments

I am tremendously indebted to Vivienne Shue, Peter Katzenstein, and Thomas Christensen for all their help. As an undergraduate at Cornell University, I took my first class on Chinese politics from Vivienne Shue, and I have been learning about China from her ever since. I can only hope that this work reflects a small fraction of her dedication and skill as a scholar. Peter Katzenstein first gave this book structure, noting something interesting at the very end of the first report I wrote back from Shanghai. I am extremely grateful for his careful reading of and insightful comments on my work. Finally, I thank Thomas Christensen for his unceasing support. Tom never seemed to doubt the value of what I was doing and never stopped giving good advice about how to improve the project.

I gratefully acknowledge financial support from the East Asia and Peace Studies programs at Cornell University. A MacArthur Foundation Fellowship in Technology and Security, a C. V. Starr Fellowship, and an L. T. Lam Award for South China Research supported the trips to China that are the backbone of this project. Within China, the people who assisted me are too numerous to acknowledge adequately here and many would prefer to remain anonymous. I thank the Shanghai Academy of Social Sciences and Tsinghua University for institutional support during my fieldwork. I spent two very rewarding years at MIT as a visiting scholar at the Center for International Studies, and I thank Richard Samuels for being so enthusiastic about my coming to Boston.

I joined the Council on Foreign Relations in October 2001 and since then have benefited immensely from Leslie Gelb's intellectual guidance and support. The council is a dynamic, challenging, and always fascinating place to work, and I thank Lawrence Korb, Elizabeth Economy, and Theophilos Gemelas for all their help and their many kindnesses. I am also grateful to Benjamin Brake for his help in editing the final manuscript.

At Cornell University Press, Roger Haydon has been an extremely able and gracious editor, and I thank him for his sage advice and enthusiasm. I especially appreciate the efforts of two anonymous readers for their generous comments on the manuscript.

It is said that your classmates determine what type of graduate school career you have, and no one has proved this to me more than Rawi Abdelal. His close readings of and insightful comments on everything that I have written have pushed me to write better and think smarter. I could not ask

for a better colleague or friend. I also greatly benefited from the cama-
raderie and input of various China reading groups: thanks to Gardner Bov-
ingdon, Rob Culp, Sara Friedman, Ben Read, Elizabeth Remick, Eric
Thun, Kellee Tsai, and Lily Tsai. Richard Locke, Barry Naughton, Judith
Reppy, Richard Samuels, Edward Steinfeld, and Richard Suttmeier all offered important comments on early drafts of the theoretical and empirical
chapters. Most of what is good and interesting in this book is due to the
people I have noted.

I cannot thank my family enough. My brother, Jonny Segal, deserves special commendation for making long trips to China and Ithaca while I was
working on this project. My parents, Freya and Anthony Segal, have remained steadfast in their love and support throughout an extremely long
journey. If I am a small, innovative enterprise, they have been the best local
government I could ask for: extremely helpful in securing access to scarce
resources but not interfering (too often) with internal business management. Richard and Cecile Sheramy have surpassed the Chinese ideal of in-
laws discussed in this book.

Finally, I dedicate this book to my wife, Rona Sheramy. Her love and
support have made it all worthwhile, and she has made me happier than I
ever believed possible.

ADAM SEGAL

New York, New York

Note on Currency

For the time period covered in this book—approximately 1980 to 2001—China had a fixed exchange rate. In the early 1980s the exchange rate ranged from RMB 2.6 to RMB 3.5 to $1. From 1986 to 1994 three different rates were used: the official rate pegged to the U.S. dollar; swap market rates, which were unofficial floating rates used by the central bank and adjusted through market intervention; and the effective exchange rate faced by exporters that was the weighted averages of the official and unofficial rates. For much of these eight years, most considered the RMB overvalued as the official exchange rate hovered around RMB 5.5. The overvaluation of the RMB acted as a de facto tax on those few technology enterprises competitive enough to export to foreign markets. Official and swap markets were unified in 1994, and the official exchange rate was set around RMB 8. The official rate was RMB 8.3 in 2000.

Digital Dragon
•••

CHAPTER ONE

High-Technology Enterprises in China

During the last several years China's technological progress has not
been insubstantial. I hope that during the 1990s, our progress will
be even quicker. Every country has to establish a clear strategic
goal, and they must achieve it. In the realm of high technology,
China must also take its place in the world.

—DENG XIAOPING,
Statements on Science and Technology, 1992

The melding of the traditional economy and information technology will provide the engine for the development of the economy
and society in the twenty-first century.

—JIANG ZEMIN,
22 August 2000

One of the main goals of the economic reform process started in 1978 by
Deng Xiaoping was to raise China's indigenous technological capabilities.
Like their counterparts in other developing countries, Chinese leaders
have been dissatisfied with and unwilling to accept their place in the international division of labor. Policymakers in these countries want to leave
behind industries dependent on low labor costs and second- or third-gen-
eration technologies, gradually replacing them with high-technology
industries. These countries are not content to remain consumers of the
newest technologies produced by more advanced economies. Industrializing states want to be as close to the cutting edge of technological innovation as possible; they hope to harness not only the economic but also the
political, military, and social benefits that are expected from technology-
intensive development. They expect, in Deng's words, to "take their place
in the world."

This book examines one set of policies adopted by China to create an indigenous technological capability: the introduction and development of
nongovernmental high-technology enterprises (*minban* or *minying keji qiye*)
in the information industries.[1] Building new technology enterprises is not

---

1. After 1993, the Chinese for "nongovernmental," *minban*, was replaced by *minying*. I have
used the later term.

easy, and many states, both developed and developing, have failed to find the right mix of policy support and market incentives. Chinese success in creating high-technology enterprises would deeply affect not only economic life within China but also the structure of the world economy and global politics.

The creation of successful high-technology firms is the next critical step for China in building a modern economy. Much of the recent work on the Chinese industrial economy has focused on either the difficulties of reforming state-owned enterprises (SOEs) or on the growth of town and village enterprises (TVEs). As the foundation of China's industrial base, SOEs warranted the scholarly attention paid to them. These enterprises provide the basic industrial inputs for the economy, the bulk of fiscal revenue for local governments, and employment and social welfare for a large part of the urban population. As in other socialist economies, the foundation these enterprises provided was somewhat shaky; they have been characterized by low productivity, declining profit, and overstaffing. Banking reform, fiscal liberalization, and overall economic growth hinged on SOE reform.

The shape of China's domestic economy, however, is changing. The share of industrial output produced by SOEs has gradually declined throughout the reforms, from 78 percent in 1978, to under 35 percent in 1995. In 2001 SOEs accounted for only 20 percent of national industrial output. The decline in industrial output was made up in the nonstate and collective sectors that have been the most dynamic parts of the Chinese economy. Employment and total factor productivity have grown much faster in the nonstate sector. TVEs now dot the Chinese countryside, and they have produced a flood of light manufactured goods for the export market. These small, flexible, market-oriented enterprises account for a large part of the growth and dynamism of the Chinese economy over the last twenty years.

That said, China's leaders do not want to remain dependent on the TVE as the engine of growth forever. Central policymakers have been growing increasingly hesitant about relying on a labor-intensive export strategy. China is eager not to miss the next wave of technological development, as it did in the 1960s, 1970s, and 1980s, and seeks to make government policy more favorable to innovation. As one Chinese commentator writes, "The greatest difficulty that all export-oriented enterprises in our country face is to produce the right products; low labor costs will not last forever. Competition between products in the international market is in fact technological competition."[2]

One of the arenas in which the Chinese hope to compete is information

2. "Heading for the World by Relying on Science and Technology," *Renmin Ribao* [People's daily], 7 December 1988, in *Foreign Broadcast Information Service-China* (hereafter FBIS-CHI), 12 December 1988, 31–32.

technologies (IT), and IT industries have grown rapidly in China. The domestic IT market was worth $168 billion in 2000, nine times larger than in
the beginning of the 1990s. During the 1990s, the IT sector registered the
fastest growth rates among the country's industrial sectors. By 2005, at the
end of the Tenth Five-Year Plan, the Chinese expect to have invested $500
billion in the sector, raising the contribution of IT to gross domestic product to 5 percent. Claims of the impact of information technologies and the
"new economy" in China have often been wildly overstated. Before the economic slowdown of 1999 and the Nasdaq market slump affected technology companies in the United States and China, many expected internet
and other IT companies to create another Silicon Valley and completely
change the way business was done on the mainland. The fizzle of internet
business in China has moderated some of the most breathless writing on
technology development, but this correction does not diminish the need
for putting the growth of technology enterprises into a larger economic
and political context. Understanding the transition from a state socialist to
a market economy during the first two decades of reform required a description of the difficult, stop-and-go process of state-owned enterprise reform. Understanding the continued transformation of the Chinese economy now demands an explanation of the process of creating an indigenous
technological capability.

The potential impact of competitive Chinese technology firms would
reach far beyond the domestic economy. There is little doubt that in the future China will have a large economy; the Chinese economy has quintu-
pled in size since the 1980s, and in 1992 the World Bank announced,
based on the purchasing parity index, the Chinese economy was the
world's third largest. But without a high-technology capability, China is unlikely to become a modern global economic power like Japan or the United
States. An indigenous technological capability would mean that, for the
first time since the Ming Dynasty (1368–1644), China would be actively involved in defining, not just accepting, international technological standards. A technologically advanced China will have less demand for foreign
technology and may slowly move away from a development strategy based
on labor-intensive goods, import substitution, and export-led growth.

Moreover, technological development addresses some of the larger
strategic concerns of the current Chinese leadership. Chinese technology
policy reveals a historically rooted concern with technological autonomy,
and an indigenous technological capability would reduce dependence on
foreign technology, especially from Japan and the United States. More directly, all modern Chinese leaders have struggled to achieve a "rich country, strong army" (*fuguo, qiangbing*), and a high technology economy would
lay the foundation for future improvements in military power, further com-
plicating Sino-American military relations. After continued tensions across
the Taiwan Straits in 1999, Jiang Zemin reportedly linked a "sound base in

technology and national defense" to the success of the mainland's reunification strategy.[3]

The process of creating nongovernmental enterprises in China has not resulted in either clear success or failure at the national level. Instead, the earliest results of attempts to create high-technology enterprises in the information industries have been uneven; variation has emerged at the regional level. At the national level, the reforms of the state science and technology system consisted of two programs: the decentralization of authority for research, development, and production to lower levels, and the creation of new spin-off enterprises defined by hybrid property rights. In the ideological context of the early reform period, the central government could not promote private enterprises and instead created the category of *minying qiye*, or nongovernmental enterprise, as a hedge between the enterprises found in centrally planned economies and the firms that operate in more open markets. But the central government never clearly defined what the nongovernmental category meant. Originally intended as a new type of property rights structure, *minying* in fact came to cover all types of property rights, including state-owned, collective, and private.

Local governments interpreted the meaning of nongovernmental differently, and so the implementation and results of technology policies varied by region. Local authorities in Beijing, Shanghai, Guangzhou, and Xi'an implemented central government directives in three areas—property rights, investment structures, and government regulation—differently, thus creating distinct local economies. Variation in policy resulted in different market structures, and so enterprise size and organization, ownership structures, and relationship of the enterprise to the state all varied regionally. These market structures were ultimately decisive for the success of high-tech nongovernmental enterprises in different regions of China. In Beijing, during the 1980s and 1990s, a number of competitive nongovernmental enterprises emerged. By contrast, during the same period, *minying* enterprises languished on the margins of the local economy in Shanghai, Xi'an, and Guangzhou.

The long-term competitiveness of the Chinese IT sector remains open to question, especially after China enters the World Trade Organization and exposes domestic producers to foreign competition. Currently many enterprises rely less on new innovations and more on copying technology from abroad; the percentage of technologies domestically produced by nongovernmental entrepreneurs is often very small. High-technology entrepreneurs still face daunting barriers to growth in China. Venture capital is scarce, property rights ill-defined, and management systems underdeveloped. Local governments have varied and continue to differ in how well they have responded to these barriers. Local officials must help entrepre-

---

3. Willy Wo-Lap Lam, "Jiang Boosts Defense Funding," *South China Morning Post*, 1 December 1999.

neurs gain access to scarce resources while not compromising enterprise autonomy.

This book explains not only how technology policies differed but also why, by placing the choices of officials in Beijing, Shanghai, Xi'an, and Guangzhou within their local contexts. Patterns of technological development emerged from the interaction among factor endowments, public policy, and local cultures. Local officials made decisions within particular economic systems; the science and technology (S&T) resources locally available, the balance of power among local actors, the locality's relationship to the central economy, and traditional patterns of industrial policy all shaped the trajectory of technological development. Seeking to develop new industrial sectors, local officials found their choices constrained by institutional resources. They also relied on traditional ideas about how to organize economic activity. These beliefs were widely shared among and provided guidance to local officials on how enterprises should be organized, how enterprises should relate to each other, and how they should interact with the local government. These economic ideas changed technological development from an apolitical outcome to a political process and helped local officials construct a narrative of development, an explanation of how they were going to move from their institutional constraints to their desired economic outcomes.

THE ROLE OF THE STATE IN TECHNOLOGICAL DEVELOPMENT

Explaining how different local patterns of development emerged is critical to understanding economic change in China. An analysis of regional variation in Chinese economic development also exposes some of the weaknesses of traditional accounts of technological development, especially in their depiction of the state. One of the central concerns of comparative political economy has been the role of the state in economic development, and the experiences of the industrializing countries of East Asia have been central to debates between free-market advocates and proponents of state intervention. Although many neoclassical economists and other scholars of development argue that the state is likely to act as an impediment to industrial transformation, extracting rents from and fostering uncertainty in domestic markets, many scholars of Japan, Korea, and Taiwan contend that the state could and in fact did play a more positive, "developmental" role. According to the latter group, central ministries such as the Ministry of International Trade and Industry (MITI) in Japan or the Economic Planning Board (EPB) in Korea played a key role by identifying and promoting critical development goals.

Chalmers Johnson, for example, argues that following World War II the Japanese central state acted like a banker, raising funds and directing the transformation of industrial sectors in cases where entrepreneurs were unable to raise the capital necessary to master the newest production technologies.[4] Late industrialization in Korea, according to Alice Amsden, required more of the state than acting like a banker. Through the allocation of subsidies, the Korean state played the role of the entrepreneur, "deciding what, when, and how much to produce."[5] Similarly in Taiwan, when local entrepreneurs did not see or were unaware of investment opportunities, the state created incentives that made productive investment decisions hard to avoid.[6] Thomas Gold argues that the Kuomintang "just did not get prices right, but it restructured society, channeled funds for investment, intervened directly in the economy, created a market plan, devised indicative plans, determined physical and psychological investment climate, and guided Taiwan's incorporation into the world capitalist system."[7]

The original terms of debate between advocates of free markets and state intervention were starker than they needed to be, and more recent scholarship has narrowed the gap between the two.[8] Though the state may have a positive role in development, it is rarely successful without the cooperation of the private sector. Different types of producer, supplier, and information networks may have a variable impact on state intervention.[9] On the other side, free-market advocates now pay greater attention to nonmarket institutions that bolster the competitiveness of individual enterprises. Development requires more than getting prices right; smoothly operating markets require clear property rights, a social welfare system, and relatively transparent administration. Government policy plays a critical role not only in opening markets up but also in fostering assets and institutions that increase competitiveness in the system.[10] The question of how (and if) the state builds market-supporting organizations is central to this book. But the Chinese case suggests that previous works of comparative political economy have focused too much on the role of the central state in creating institutions at the national level. For example, the success of developmental states, according to the sociologist Peter Evans, depends on a balance between institutional autonomy and dense links to societal actors. Informal

4. Chalmers Johnson, *MITI and the Japanese Miracle: The Growth of Industrial Policy, 1925–1975* (Stanford: Stanford University Press, 1982), 27–28.

5. Alice Amsden, *Asia's Next Giant: South Korea and Late Industrialization* (New York: Oxford University Press, 1989), 143.

6. Robert Wade, *Governing the Market: Economic Theory and the Role of the Government in East Asian Industrialization* (Princeton, N.J.: Princeton University Press, 1990).

7. Thomas Gold, *State and Society in the Taiwan Miracle* (Armonk, N.Y.: M.E. Sharpe, 1986), 122.

8. The emerging consensus on state-market relations is reflected in World Bank publications like Shahik Burki and Guillermo Perry, *Institutions Matter: Beyond the Washington Consensus* (Washington, D.C.: World Bank Latin American and Caribbean Studies, 1998).

9. Chung-in Moon and Rashemi Prasad, "Networks, Politics, and Institutions," in *Beyond the Developmental State: East Asian Political Economies Reconsidered,* ed. Steve Chan, Cal Clark, and Danny Lam (New York: Palgrave, 1998), 9–24.

10. David G. McKendrick, Richard F. Doner, and Stephan Haggard, *From Silicon Valley to Singapore: Location and Competitive Advantage in the Hard Disk Drive Industry* (Stanford, Calif.: Stanford University Press, 2000).

networks, both internal and external, give MITI or the EPB an internal coherence and corporate identity, and also tie officials to private power holders.[11] Bureaucratic actors may be autonomous and coherent, but they are also embedded in a concrete set of social relations. Too much autonomy distances governments from social actors and makes them dependent on decentralized private actors for implementation. Too little and state actors, subject to individual or special interest group demands, cannot resolve collective action problems. State capacity and the ability to transform the economy is a product of what Evans calls "embedded autonomy": bureaucratic insulation with intense connections to the surrounding social structure.[12]

Unlike in Japan or Korea, no organ of the central state in China has been able to intervene effectively in new technology markets. The Chinese central state lacks both internal coherence and strong ties to society.[13] Moreover, the state, at least at the center, has not been able to overcome these institutional weaknesses by developing expansive ties to important social actors.[14] The Ministry of Science and Technology, which replaced the State Science and Technology Commission (SSTC) and was the government bureau most responsible for nongovernmental enterprises, lacked administrative power. Moreover, *minying* entrepreneurs rarely speak of the Ministry of Information Industries (MII) or any other ministry as critical to their development. Central state ministries like the MII are more likely to be seen as barriers to growth, restricting the access of foreign investors to domestic producers and tightly controlling content on the internet.

The state still plays a critical role in technological development in China, that role is just not limited to the highly visible and identifiable organs of the central state that have been the main subject of state-centered approaches. The Chinese state is a less monolithic, much more diffusely dispersed set of institutions than suggested by Evans and by the developmental state literature. Government support remains crucial to growth in China, but the most important links are to the enterprise's supervisory agency (*zhuguan danwei*), which can be a university, research institute, state-owned enterprise, or local government. All of these are "state" actors, but none resemble MITI or the EPB. The Chinese case forces us to view the state as a more decentralized organizational structure embedded within a range of institutional, political, and social arrangements at the central and local level.

11. Peter Evans, *Embedded Autonomy: States and Industrial Transformation* (Princeton, N.J.: Princeton University Press, 1995), 49.

12. Ibid., 50.

13. Michael Oksenberg and Kenneth Lieberthal, *Policymaking in China: Leaders, Structures, and Process* (Princeton, N.J.: Princeton University Press, 1988).

14. Vivienne Shue, "State Power and Social Organization in China," in *State Power and Social Forces: Domination and Transformation in the Third World*, ed. Joel Migdal, Atul Kohli, and Vivienne Shue (Cambridge: Cambridge University Press, 1994), 67.

Instead of discussing the role of the state, this book focuses on the behavior of and the interaction between the central and various local governments involved in technology development. The "state" is used to refer to a set of political relations that encompasses branches of the central, provincial, city, and district governments. These relations define politics within the state, between the state and society, and between the polity and some international actors.[15] Viewing the state as a set of relations makes it easier to see development as the outcome of both the competition and interaction between the central and local governments. The decentralization of state authority to lower levels has increased the power and autonomy of local governments, but the local and central governments are not polar opposites, engaged in a zero-sum game. Local governments operate in a context shaped by local institutions and the central government, and the two levels of government reinforce each other.

The case of technology policy in China also demonstrates that the process of economic development is not simply the story of the central state intervening in a cohesive national economy. The national economy needs to be disaggregated like the central state. Geographically distinct economies can exist within national boundaries. Richard Locke argues that distinct regional economies continue to exist within the seemingly unified Italian national economy.[16] And Gary Herrigel notes, "Different actors, in different regions, with different histories, institutions, and cultures, will invariably conceptualize, organize and enact industrial activity in ways that reflect their own peculiar pasts and contextually and discursively engendered conceptions of and strategies for the future."[17] Yet in most explanations of economic development, the subnational level disappears; growth is the result of creating the appropriate political and economic institutions at the national level.[18] While regional economies may exist within national economies, they are generally portrayed as premodern remnants to be gradually absorbed into the central economy.

In fact, the process of development occurs at both the national and subnational level and focusing solely on the national level overlooks many of the mechanisms that are actually driving development.[19] Herrigel describes

15. Peter J. Katzenstein, *Cultural Norms and National Security: Police and Military in Postwar Japan* (Ithaca, N.Y.: Cornell University Press, 1996), 4. Also see Vivienne Shue, *The Reach of the State: Sketches of the Chinese Body Politic* (Stanford, Calif.: Stanford University Press, 1988), 4.

16. Richard Locke, *Remaking the Italian Economy* (Ithaca, N.Y.: Cornell University Press, 1995), 24–27.

17. Gary Herrigel, *Industrial Constructions: The Sources of German Industrial Development* (Cambridge: Cambridge University Press, 1996), 26.

18. These approaches were influenced by the national model school developed by Andrew Shonfield. See his *Modern Capitalism: The Changing Balance of Public and Private Power* (New York: Oxford University Press, 1965). See also the authors collected in Peter J. Katzenstein, ed., *Between Power and Plenty: Foreign Economic Policies of Advanced Industrial States* (Madison: University of Wisconsin Press, 1978).

19. Adam Segal and Eric Thun, "Thinking Globally, Acting Locally: Local Governments, Industrial Sectors, and Development in China," *Politics & Society* 29, 4 (December 2001): 557–88.

how in Baden Württemberg small and large firms have developed a regionally based comparative advantage lacking in other parts of Germany. The
regional economy has developed faster than other parts of the national
economy by "socializing risk across a broad array of public and private organizations."[20] Similarly, AnnaLee Saxenian notes that in high-technology
sectors, firms in Silicon Valley created a regional economy better able to respond to market changes in the late 1980s than corporations around
Route 128 in Boston.[21] Dense social networks and open labor markets characterized Silicon Valley's regionally based industrial system and promoted
collective learning, experimentation, and entrepreneurship.

Regional economies continue to play an important role in developed
economies, and their influence may be even more critical and widespread
in more transitional ones. Countries such as China that are undergoing
rapid economic and social change have less uniformity than countries that
have enjoyed long periods of stability. Even in developing countries with
more centralized government institutions than China, pockets of development, or "industrial clusters," have emerged within very low rates of overall
national growth. Horizontal relations that foster information exchanges,
skill transfers, and capital accumulation characterize these clusters.[22]

Economic reform in China has produced a national economy that looks
like a mosaic of regional economies.[23] The center expanded both the decision-making authority of local governments and their ability to retain the
revenue earned within their respective jurisdictions.[24] As a result, localities
pushed local development, sometimes by implementing a local industrial
policy that ignored national objectives. Even on issues officially controlled
by the center, a local government's interpretation and its degree of compliance with the directives were often the more important determinants of
actual policy.

The developmental state approaches discussed above offer important
insights into technological innovation by focusing attention on the critical
role played by state actors and the policies that created the organizational

20. Gary Herrigel, "Large Firms, Small Firms, and the Governance of Flexible Specialization: The Case of Baden Wurttemberg and Socialized Risk," in *Country Competitiveness*, ed. Bruce
Kogut (New York: Oxford University Press, 1993), 17.

21. AnnaLee Saxenian, *Regional Advantage: Culture and Competition in Silicon Valley and Route
128* (Cambridge, Mass.: Harvard University Press, 1994), 2–3; and AnnaLee Saxenian, "Regional Networks and the Resurgence of Silicon Valley," *California Management Review* 33, 1
(1990): 89–113.

22. Khalid Nadvi and Hubert Schmitz, *Industrial Clusters in Less Developed Countries: Review of
Experience and Research Agenda,* Institute of Development Studies, Discussion Paper 339 (January
1994), 36.

23. It is probably more accurate to say that reforms were the catalyst for the reemergence of
local economies; regional economies existed in China even under the central plan. See G. W.
Skinner, "Marketing and Social Structure in Rural China," *Journal of Asian Studies* 24, November
1964, February 1965, May 1965.

24. Susan L. Shirk, *The Political Logic of Economic Reform in China* (Berkeley: University of California Press, 1993).

space for entrepreneurial activity. Yet we can only understand the process of development by understanding the decisions local actors made, and how these actors were embedded in distinct local histories with different institutional constraints and opportunities.

ONE CHINA, MANY ECONOMIES

The growing importance of local governments has been widely documented in post-Mao China. Jean Oi, for instance, argues that increased fiscal incentives gave rise to a new form of state-led growth in rural China, what she calls local state corporatism: a system in which local governments "treat enterprises within their administrative purview as one component of a larger corporate whole."[25] Victor Nee focuses on how increased market pressure led local governments and private firms to form alliances as protection against an uncertain environment, and Susan Whiting explains why property rights arrangements evolve in different ways in different regions and how this variation affects the extractive capacity of the state.[26]

Andrew Walder's work on town and village enterprises is particularly helpful for our thinking about different types of local governments and technological development. In explaining variation in rates of growth between dynamic TVEs and seemingly moribund SOEs, Walder does not focus on property rights alone and instead notes important institutional differences between government actors in rural and urban contexts.[27] Industrial productivity and growth vary between the two areas because the organizational characteristics of local governments vary, not because some firms are collectively owned, while others are owned by the state. The size and degree of internal diversification of local governments affect the intensity of financial incentives and budget constraints. Lower levels of government mainly located in rural areas can more effectively monitor enterprises, enforce financial constraints, and are less likely to burden enterprises with nonfinancial objectives. Governments at higher levels face greater political constraints, larger nonfinancial burdens, and a limited ability to monitor enterprise performance and enforce financial discipline. In the case of IT, there was not a significant difference between the size and diversification of the local governments discussed in this book. Other organizational characteristics—the balance of power among different gov-

25. Jean Oi, "Fiscal Reforms and the Economic Foundations of Local State Corporatism in China," *World Politics* 45, 1 (October 1992): 99–126.

26. Victor Nee, "Organizational Dynamics of Market Transition: Hybrid Property Forms and Mixed Economy in China," *Administrative Science Quarterly* 37 (1992): 1–27; Susan Whiting, *Power and Wealth in Rural China: The Political Economy of Institutional Change* (Cambridge: Cambridge University Press, 2001).

27. Andrew Walder, "Local Governments as Industrial Firms: An Organizational Analysis of China's Transitional Economy," *American Journal of Sociology* 101, 2 (September 1995): 263–301.

ernment departments at the local level, the relationship of the locality to the center—still differed in ways that affected technological development.

What was strange in the case of technological development was not that institutional structures varied within China, but that the institutional structures most compatible with enterprise development were in Beijing. Highly centralized S&T systems, like the one found in China, tend to stifle innovation.[28] State policy usually has a pervasive and pernicious influence on firm organization, encouraging centralized and hierarchical organizations inappropriate to rapid technological development. Ties to the central government, easily developed though geographical proximity and reliance on the government as a primary customer, often encourage firms located in the capital to replicate the centralized organization of the state. Guaranteed markets, development aid, and subsidies all diminish market pressures for organizational changes within firms. Few countries boast of an innovative technology center located in the political capital; in India software firms thrive in Bangalore, in France innovative firms have emerged near Grenoble, in the United Kingdom outside Cambridge, in Italy outside Milan, and in the United States in Boston and northern California.

Given these difficulties, we would expect high-tech enterprises in Beijing to be less able to re-create themselves independently of the central government than enterprises in other parts of the country. In fact, through building housing and providing other social benefits, some high-tech enterprises have re-created the institutional features of the *danwei,* the work unit at the center of state-owned enterprises.[29] The growing regionalization of the Chinese economy should also increase the incentives for locating high-tech enterprises outside of the center. During the 1980s and 1990s the highest rates of growth were in coastal provinces like Fujian, Guangdong, and Jiangsu. In these areas, local officials and managers of collective enterprises were the most active in developing extensive networks to help support industrial development.[30] If a locality can mobilize local S&T resources (or attract them from other provinces), high-tech enterprises, with their high rates of profit and widespread impact on the local economy, should represent an attractive potential development ally for local governments.

Added to the growth of Beijing, the lack of IT entrepreneurship in Shanghai was itself counterintuitive. At the beginning of the reform process, there were a number of reasons to believe that its technology enterprise development would be more robust than Beijing's. First, Shanghai

28. Robert Gilpin, *France in the Age of the Scientific State* (Princeton, N.J.: Princeton University Press, 1968); and John Zysman, *Political Strategies for Industrial Order: State, Market, and Industry in France* (Ithaca, N.Y.: Cornell University Press, 1977), 19.

29. Corinna-Barbara Francis, "Reproduction of the Danwei Institutional Features in the Context of China's Market Economy: The Case of Haidian District's High-Tech Sector," *China Quarterly* 147 (September 1996): 839–59.

30. You-tien Hsing, *Making Capitalism in China: The Taiwan Connection* (New York: Oxford University Press, 1998), 118.

was both the country's most important supplier of electronics products and a key research and development (R&D) base. Second, Shanghai already played a leading role in the consumer electronics sector. In 1981, for example, Shanghai produced 22 percent of all televisions and 37 percent of all tape recorders in China.[31] Finally, even cultural stereotypes favored Shanghai: the Shanghainese are usually viewed as entrepreneurial business people, while their Beijing counterparts are viewed as candid peasants more attuned to politics than business.[32]

To explain these outcomes, I build on the work on local governments done by Oi, Nee, Walder, and others. Their studies, however, tend to focus on one region and one type of industry. Local state corporatism may be an effective developmental strategy, but it may not work under all conditions for all types of industrial sectors. Institutional arrangements need to be compared across similar administrative units, not just between municipal governments and rural townships. By looking at the same sector across China, and at how development patterns varied by region, we can better link local institutions with the failure or success of specific sectors. Understanding the development patterns of Beijing, Shanghai, Guangzhou, and Xi'an forces us to focus more closely on how the capacity of state institutions and economic organizations at the local level determine a region's developmental capacity.

WANTED: A GOOD MOTHER-IN-LAW

Macro-level state policies created the category of nongovernmental enterprise, but the development trajectories of these enterprises depended on the decisions of local government officials. The choice for a local government in China was not whether it should intervene, but how to intervene and best help organize new industrial sectors. And some local governments made wiser choices than others, supporting enterprise systems more compatible with the needs of the information industries.

In some industries, it is not uncommon for a local government to be directly involved in production. The local government acts as an important monitoring agency, settling disputes between suppliers; the local government may help overcome collective action problems or even own shares of production facilities.[33] When production is dispersed among diverse groups

31. Detlef Rehn, "Organizational Reforms and Technology Change in the Electronics Industry: The Case of Shanghai," in *Science and Technology in Post-Mao China*, ed. Denis Fred Simon and Merle Goldstein (Cambridge, Mass.: Harvard University Press, 1989), 144.
32. Yang Dongping, *Chengshi Jifeng: Beijing he Shanghai de Wenhua Jingshen* [City monsoon: The cultural spirit of Beijing and Shanghai] (Beijing: Dongfang Chubanshe, 1994).
33. Mark Granovetter, "Economic Action and Social Structure: The Problem of 'Embeddedness,'" *American Journal of Sociology* 91, 3 (1985): 481–510; and Mark Granovetter, "Economic Institutions as Social Constructions: A Framework for Analysis," *Acta Sociologica* 35, 1 (1992): 3–13.

of producers and suppliers within a region, local governments can provide specific administrative functions that support the region's production system.[34]

Even in sectors where the authorities are less centrally involved in production networks, local governments can shape new markets in other ways. In particular, governance structures, the social institutions that guide the interaction between enterprises and between enterprises and the local government are especially susceptible to government influence at the beginning stages of market creation. As the sociologist Neil Fligstein argues, "Property rights, governance structures, and rules of exchange are arenas in which modern states establish rules for economic actors. States provide stable and reliable conditions under which firms organize, compete, cooperate, and exchange."[35] Uncertainty in external markets means that technology firms often develop extensive social ties to other firms within the same geographical region. Innovation involves multiple actors and springs from the combination of various actors with specialized and complimentary competency and knowledge.[36] Social, political, and economic organizations located outside the enterprise can bolster internal capabilities. Links among various actors—formal or informal, institutionalized in written contract or informal trade practices—facilitate information flows, allow for specialization, mediate conflict, and define the terms of cooperation and competition between firms.[37]

The fact that many of these institutions are civic organizations or producer associations has led some to overstate the degree to which they are independent from government actors. This has been especially true in descriptions of Silicon Valley, where the roles of the federal and California governments have attracted little discussion.[38] Moreover, the belief that technology sectors change too rapidly for effective government intervention is fairly widespread. Yet there are at least two reasons to move government actors back to the center of technology development, one more general, the other specific to developing economies. First, even in the more developed economies, nonmarket institutions that support the high-tech market are in fact heavily influenced by government actions.[39] As Lawrence Lessig notes, "Innovation has always depended on a certain kind of regula-

34. For the role that Shanghai's local government plays in the auto industry, see Eric Thun, "Changing Lanes in China: Reform and Development in a Transitional Economy" (Ph.D. diss., Harvard University, 1999).

35. Neil Fligstein, "Markets as Politics: A Political-Cultural Approach to Market Institutions," *American Sociological Review* 61 (1996): 660.

36. OECD Proceedings, *Boosting Innovation: The Cluster Approach* (Paris: OECD, 1998), 11.

37. Gerald Davis and Heinrich R. Greve, "Corporate Elite Networks and Governance Changes in the 1980s," *American Journal of Sociology* 103, 1 (July 1997): 1–38; and G. B. Richardson, "The Organization of Industry," *Economic Journal* 82, 327 (September 1972): 883–96.

38. Saxenian, *Regional Advantage;* and the reviews of *Regional Advantage* in *Economic Geography* 71, 2 (April 1995): 199–207.

39. Bai Gao, *Economic Ideology and Japanese Industrial Policy: Developmentalism from 1931 to 1965* (New York: Cambridge University Press, 1997), 7.

tion."[40] Government decisions about regulation, property rights, licensing, and standards all influence the structure of social institutions.

Second, without some sort of government support, the chances of a technology enterprise succeeding in China were exceedingly slim. There are distinct disadvantages to being a latecomer in technological development, and unlike in neoclassical accounts of development, the market alone could not overcome problems like incomplete information.[41] At the early stages of development in China, civic or producer groups did not exist to foster ties between technology enterprises. Ties to public institutions were essential to the early stages of growth, especially given China's small technological market and its lack of venture capital.[42] As one manager in a (successful) Xi'an enterprise bluntly put it: "No government support, no nongovernmental enterprise success."[43]

This does not give local officials justification to intervene whenever and however they want. Local governments had to provide a certain type of support. There were more and less successful strategies of government intervention that fell somewhere in between the spectrum of market-driven and state-led development strategies. In the case of Taiwan and Korea, Stephan Haggard describes a set of interventions that reduced the risks of shifting into the export business for domestic firms by providing various incentives and by lowering information and transaction costs.[44] First, Taiwan and Korea used a strategy of import substitution (IS) to build up a market in intermediate and capital goods and to strengthen domestic manufacturing capabilities. Soon after, this strategy was replaced with the liberalization of imports and exchange rate reform. Key to the success of Taiwan and Korea was that these interventions did not last too long. Taiwan and Korea did not engage in IS long enough to distort the incentives for domestic producers or create opportunities for government actors to engage in rent seeking. Once enterprises upgraded their manufacturing capabilities they were exposed to international pressure; infant industries had to move quickly into the global market.

With Chinese information industries, successful government intervention has been distinguished less by chronology and more by scope. That is, most of the earliest nongovernmental enterprises entered areas where domestic producers already had a comparative advantage; even if the domestic market had not been protected, there were few, if any, foreign firms that

40. Lawrence Lessig, "Innovation, Regulation and the Internet," *The American Prospect* 11 (27 March 2000). Available online at http://www.prospect.org/print/V11/10/lessig-l.html.

41. Alexander Gerschenkron, *Economic Backwardness in Historical Perspective* (Cambridge, Mass.: Harvard University Press, 1962).

42. Scott Kennedy argues that *minying* firms actually succeed because they have no ties to the government. See "The Stone Group: State Client or Market Pathbreaker," *China Quarterly* 152 (December 1997): 746–77.

43. Interview, no. X8, 27 July 1998.

44. Stephan Haggard, *Pathways from the Periphery: The Politics of Growth in the Newly Industrializing Countries* (Ithaca, N.Y.: Cornell University Press, 1990), 93.

had the technology and skills necessary to produce Chinese language software.[45] In the Chinese case, local government support helped enterprises overcome many of the barriers that characterize new markets: incomplete information, undefined property rights, and price distortions.

Moreover, and somewhat paradoxically, government intervention could protect domestic producers from extractive state agencies. Local commercial and industrial bureaus, for example, were less likely to levy illegal taxes on enterprises linked to other branches of the local government. In addition, entrepreneurs could rely on these government actors to settle disputes with other public agencies. For technology enterprises to succeed, local governments had to balance the desire to help with the tendency to infringe on enterprise autonomy. This was not easy to do since the lines of investment and ownership and managerial authority were not clearly defined or demarcated at the beginning stages of reform. Entrepreneurs often complained that supervisory agencies confused making a loan with investing.

Entrepreneurs referred to the ideal type of support as being a "good mother-in-law" (*hao po po*). The best mother-in-law did not interfere with the internal workings of the enterprise, with the relationship between husband and wife, but supported the couple in their search for a new apartment or the raising of the children. Not having a "mother-in-law" meant the enterprise was on its own, unable to influence official agencies and susceptible to a range of extractive agents. Entrepreneurs wanted to maintain stable relations with one supervisory agency, and only one. As one entrepreneur put it, "Having no mother-in-law is bad, having too many is even worse."

LOCAL GOVERNMENTS AND TECHNOLOGY POLICY

Government decisions about property rights, funding structures, and how (and how often) to supervise enterprises significantly shaped enterprise structures. These decisions and other government pronouncements also sent more indirect signals to entrepreneurs about what type of enterprises would or would not be accepted or supported in an uncertain political environment (see table 1.1). Beijing combined guidance to entrepreneurs while still allowing them to shape the emergence of the sector. Local officials in Xi'an and Shanghai tended to provide government support while at the same time interfering with internal enterprise management. Guangzhou fell toward the other extreme; government officials did not intervene in daily business operations, but they also did not provide the supervision required to overcome market failures.

45. This is the case for domestic competitors as well. The first nongovernmental enterprises entered new sectors where there was no SOE presence and no ministerial supervision.

Table 1.1. Types of Local Government Support

	Beijing	Shanghai	Guangzhou	Xi'an
I) Investment	Science based, research institutes	Labs within SOEs	Basic infrastructure	Science based, research institutes
	Loans, FDI to *minying* enterprises	Loans, FDI funneled to SOEs	Only FDI to small collective private enterprises	Loans to SOEs, very limited FDI
II) Property Rights	Multiple forms of ownership	Limited stock companies for SOEs; large groups (*jituan*)	Collective then private	SOEs, limited stock companies
III) Government Supervision				
a) Market Activities	Horizontal	Vertical ties; large groups	Hands off	Horizontal and vertical
b) Political Activities	*Minying* enterprises at center of technological development	*Minying* as complement to SOEs	All private	Between being at center and being a complement to SOEs

Local governments had a surprising degree of freedom to determine ownership structures. The arrival of the first *minying* enterprises in the early 1980s forced local governments to decide when they were going to recognize these enterprises and create procedures for registering and regulating enterprises. After it became clear that ambiguous connections between enterprises and supervisory agencies limited growth, some officials encouraged collective enterprises to register as private or nongovernmental enterprises. Some officials, however, moved in the opposite direction, registering private enterprises as collective; this was known as "wearing the red hat" (*dai hong maozi*) and protected local officials from attacks that they were encouraging still-suspect private ownership. Later local governments had to decide when and if they were going to encourage the formation of limited stock companies, the issuance of technology shares, and the listing of companies on domestic and foreign stock markets. These structures affected the distribution of wealth within enterprises and the incentives entrepreneurs and scientists have for engaging in productive activity.[46] The

46. Margaret Blair, *Ownership and Control: Rethinking Corporate Governance for the Twenty-first Century* (Washington, D.C.: Brookings, 1995), 11; and David Soskice, "German Technology Policy, Innovation, and National Institutional Framework," *Industry and Innovation* 4, 1 (1997): 77.

most efficient *minying* enterprises created a system that allocated authority to the individuals most likely to have the information necessary to use resources efficiently to create wealth.

Local government decisions about funding were equally important to new organizational structures. The most important signal that *minying* enterprises were to be encouraged was the expansion of bank loans to enterprises; officials could either direct banks to loan to nongovernmental enterprises, act as guarantors for enterprises seeking loans, or create innovation funds that directly loaned to enterprises. Local governments could also influence enterprise structure by funneling R&D expenditures to independent research institutes as opposed to labs attached to SOEs, or by deciding to fund the development of new technologies instead of raising the level of current equipment.

Finally, local governments had to decide how they were going to supervise nongovernmental enterprises. The most activist governments maintained a position on the board of directors. Other local governments limited their interventions into the business operations of the enterprise, but adopted more diverse social and political roles in the new sector: supporting the *minying* enterprise association, assisting with the acquisition of residence permits (*hukou*) for S&T personnel from outside the city, or arranging management seminars for new entrepreneurs. In addition, local governments often created new organizations that fostered horizontal links between research and production units or allowed scientists to moonlight from their public-sector jobs for private or collective enterprises. These actions often encouraged scientists to leave their research jobs and establish their own enterprises.[47]

PATTERNS OF DEVELOPMENT

The interaction of national technology policy and local governments created distinct patterns of nongovernmental enterprise development that differ along three dimensions: the quantity and quality of local government ties, enterprise size and ownership structures, and the extent of horizontal ties to other actors (see table 1.2). These dimensions were important for the types of IT most likely to develop in a locality. In short, these development patterns fit certain types of enterprises and certain types of technologies.

Even within the broad category of information technologies, specific types of technologies require different types of production, coordination, and regulation. The size of initial capital investments, the degree of vertical integration, and governance within and between enterprises may all vary

47. This most closely resembles the roles of "midwife" and "husbandry" as described by Evans in *Embedded Autonomy*, 13–14.

Table 1.2. Patterns of Enterprise Development

	Beijing	Shanghai	Xi'an	Guangzhou
Ownership Structure	Hybrid	State-owned	State-owned and hybrid	Private
Enterprise Size	Mainly small	Large	Mainly large	Small
Managerial Authority	Founding individuals	Bureaucrats	Founding individuals, bureaucrats	Founding individuals
Type of State Contact	Financial support + supervision	Financial support + meddling	Financial support + meddling	Limited financial support + no meddling

based on the needs of the technological sector. Software producers, for example, are generally independent enterprises that compete with other developers for investment. Investment does not have to be all at once, but can come in distinct stages. Enterprises often adopt a "flat" organizational structure, trying to reduce the barriers between the market and the enterprise, and they often rely on horizontal and informal ties to other enterprises to provide access to highly specialized but complementary knowledge about production, markets, and innovation. In contrast, semiconductor manufacturing is more like a traditional industry. Innovation in the sector has slowed and manufacturing requires huge initial investments, a high degree of vertical integration, and a hierarchical organizational structure.

Enterprises in Beijing were linked to a wider range of supervisory agencies at all levels of government, but they have gradually become more autonomous of the central and local government than their counterparts in Shanghai. The supervisory agency officially in charge of the enterprise was in the minority on the board, and the percentage of profits the agency can claim was shrinking. Individual managers began management reforms, diffusing product authority through the enterprise and encouraging relationships between related tasks. Moreover, dense social horizontal networks linked enterprises to each other and to local institutions, encouraging entrepreneurship and experimentation.

As a result, nongovernmental enterprises were at the center of Beijing's growth. By 1993 *minying* enterprises in Beijing made up 10.2 percent of total output value in Beijing. Beijing enterprises were responsible for RMB 6.75 billion in income, 84 percent of the national total generated by nongovernmental enterprises.[48] Moreover, in 1996, the gross income of the

48. Beijingshi Kexue Jishu Weiyuan Hui, *1997 Niandu Beijing Keji Qiye Gongzuo Yaoloan* [1997 overview of Beijing technology enterprises] (Beijing: Zhongguo Jingji Chubanshe, 1998), 70.

total 6,787 *minying* enterprises in all of Shanghai totaled over RMB 17.6 billion;[49] in Beijing the 4,506 in the zone alone broke RMB 30 billion. Including all *minying* enterprises in Beijing would push the total up to RMB 45 billion.[50]

From 1978 to 1990 over 50 percent of award-winning innovations were based on interorganizational collaboration. The largest type of collaboration within this category was between a research institute and a business enterprise. Legend, Founders, and Kehai would all be classified as this type of enterprise, have all received recognition from local and national science commissions, and all are located in Beijing. In 1988, Legend received a National Award for Scientific and Technological Progress for its Chinese character system; in 1992 its personal computer won the same award.[51] From 1988 to 1994, Beijing nongovernmental enterprises received 32.5 percent of all awards given to national technology entrepreneurs, more than in any other part of the country.[52] In addition, in 1994 almost two and a half times more patents were issued in Beijing than in Shanghai.

This pattern of development has been conducive to a certain type of IT development. Beijing led the country in software exports from 1996 to 2001 and the capital led in internet start-ups as well. In July 2000, Shanghai had only 8,457 of the mainland's domain-name registrations, 8.6 percent of the whole country, and less than a quarter of the numbered registered in Beijing. The number of internet users in Shanghai was less than 60 percent of Beijing's web population.

Shanghai was dominated by a small number of larger enterprise groups, fragmenting technological networks. Local governments were often able to influence business operations by having a representative on the enterprise board of directors. These enterprises were supposed to have more flexibility in management and market decisions, but the question remains if that was possible with local government representation on management boards. Within the enterprises themselves, authority was highly centralized. These enterprises had few horizontal ties to each other and little conception of themselves as independent actors leading a technological revolution; they remained at the margin of the public-sector economy. Managers were often senior bureaucrats, technological development typically was government directed, and as a consequence enterprises were unable to respond to the rapid changes in the market.

Small, nongovernmental enterprises play a limited role in Shanghai's technology sector. The six largest state-owned enterprise groups generated

---

49. "Bai jia Minke Qiye Huo Biao Zhang" [Various nongovernmental science enterprises receive praise], *Shanghai Keji Bao* [Shanghai science and technology daily], 22 May 1996, 1.
50. "Shi Yan Qu Gongbu 1996 Jingji Baipishu" [Experimental zone publishes 1996 economic white paper], *Beijing Keji Bao* [Beijing science and technology daily], 17 February 1997.
51. "Lianxiang Jituan" [Prospectus for Legend group], n.d., 6.
52. "Beijing Minying Keji de Chenggong Moshi" [Patterns of Beijing nongovernmental firm's success], *Zhongguo Keji Chanye Yuekan* [Chinese technology industry monthly] 9 (1995): 27–28.

87 percent of total output value in the IT sector in 2000.[53] Shanghai has become the country's largest semiconductor producer and leads in the manufacturing of IT machinery and hardware. Although Shanghai lags in internet start-ups, the local government has played an aggressive role in promoting broadband. Shanghai has emerged as a broadband hub with the most advanced infrastructure in the country.

Xi'an exists somewhere between Shanghai and Beijing. Companies tended to be small, with decentralized management structures and few barriers between marketing, research, and sales. Xi'ian's large concentration of universities and research institutes meant that enterprises focused on R&D-intensive development. But the lack of foreign capital meant the most successful enterprises have high product diversity. While older *minying* enterprises in Beijing gradually formalized their relations with and distanced themselves from their supervisory agencies, enterprises in Xi'an sought closer relations with local authorities. As they grew larger, they attracted the attention of the authorities and formed relations with various branches of the local government.

At the farthest end of the spectrum, enterprises in Guangzhou also tended to be small, but with almost no support from the state. In Guangzhou, funding was funneled away from R&D to more profitable sectors, and cooperative relationships developed through joint ventures. As a result, enterprises focused on the quick profits available in buying and selling already established technologies (often imported), rather than face the risks and uncertainty of developing new technologies.

METHODOLOGY AND CHAPTER OVERVIEW

In this book I examine four cases of high-technology development in one industrial sector: the information industries. Most nongovernmental high-tech enterprises are involved in information technologies, and the sector includes enterprises involved in personal computer, component, and peripherals production, software development, and internet services. High entry cost, high risk, and the continued dominance of central state research labs and factories mean there are few if any private entrepreneurs involved in the development of other technology areas like semiconductors, new materials, or new energy sources.[54]

IT is an extremely useful lens through which to focus questions about government intervention into new markets.[55] Information industries have influenced production processes in all sectors and make up a growing

53. "Vice Mayor: Shanghai Leads PRC IT Industry," *Xinhua*, 9 May 2000, in FBIS-CHI, 9 May 2000.

54. See Wu Xijun, *Gaojishu: Kuashijie de Zhanlue Wenti* [High technology: The next century's strategic problem] (Jiangsu: Jiangsu Kexue Jishu Chubanshe, 1992).

55. Evans, *Embedded Autonomy*.

share of output in all advanced economies. Moreover, IT is a hard case for any arguments about government intervention. Rapid change, high capital investments, and high risk all make IT an especially difficult challenge, and policymakers, in both the developed and developing world, are increasingly concerned with their international competitiveness in these sectors.

Primary-source information for this book came from more than 120 interviews with local government officials and high-technology entrepreneurs in the IT sector in Beijing, Shanghai, Guangzhou, and Xi'an.[56] As further context for these interviews, I consulted planning documents, statistical yearbooks, newspaper reports, and articles in specialized journals. As mentioned earlier, the development patterns and success of *minying* enterprises have varied by region, and the cases were chosen to give the full variation on the dependent variable. All four cities wanted to make IT a "pillar industry" (*zhizhu chanye*) of the economy. Shanghai and Beijing both have highly developed S&T resources but adopted different policies toward high-technology enterprises, with the result that *minying* enterprise development lagged in Shanghai. Guangzhou and Xi'an lacked the same resources and turned their attention to nongovernmental enterprises later in the reform process. In addition, local officials in these two cities had to balance municipal development with provincial political concerns. Shanghai and Beijing are both centrally planned cities (*jihua danlieshi*), officially administered by the center; officials in the provinces had different relations with the central government.

The lesson of the Chinese case is that under conditions of political and market uncertainty local governments must provide a certain type of help for technological development. Government intervention is frequently intrusive, but sometimes it can be productive. Without well-developed industrial associations and clear property rights, government actors can bolster individual enterprise capabilities. Local governments can help build non-market institutions that still allow high-technology entrepreneurs to shape the sector. These actions do not substitute for infrastructure development and the consistent enforcement of laws, but government action can be tailored to meet the needs of new enterprises. The Beijing local government was more successful in providing access to capital and technology while maintaining enterprise autonomy. By contrast, Shanghai and Xi'an were more likely to provide too much help, securing loans for large SOEs while interfering with R&D strategies and business operations. Finally, Guangzhou did not provide enough support, ignoring the need of fledgling enterprises for financial assistance or property rights clarification.

If the actions of local governments explain China's variable success in creating high-technology enterprises, what explains the different actions of local governments? After describing the policies that created nongovernmental enterprises, I take up this question in chapter 2. In order to under-

---

56. Interviews are listed in the appendix.

stand why local governments adopted their strategies, technology policy must be situated within the local political, economic, and social context. This context includes factor endowments, the nature of the links between government and societal actors, and the role of the locality in the national economy. All possibilities were not available to local governments; path dependence played an important role in limiting the development options open to local actors.

Inherited industrial structure can only point to the direction local governments may take, not to the specific path followed. Geography is not destiny, and a material explanation alone is apolitical. Technology policy was made within distinct cultural contexts, shaped by shared beliefs that defined the goals and purposes of technological development. How local governments approached the process of *minying* enterprise creation—which development path they chose to follow—was a political choice. Chapter 2 concludes by examining the implications of local institutional constraints to the larger question of economic transition in China and in the former communist countries of Eastern Europe.

In the next three chapters, the politics of *minying* enterprise development in Beijing, Shanghai, Guangzhou, and Xi'an are described. Chapter 3 focuses on Beijing and its definition of nongovernmental enterprise as small, flexible, and relatively autonomous; local policies tried to institutionalize this definition as broadly as possible. Chapter 4 shows Shanghai much more concerned with the role of state-owned enterprises. *Minying* enterprises were expected to complement the public sector, and local policies provide little industry-specific support for small enterprises. The cases of Guangzhou and Xi'an are contrasted in chapter 5. Both cases started later than Shanghai and Beijing, both cases faced significant disadvantages in terms of factor endowments. In Guangzhou, local officials were more concerned with defining enterprises as private and attracting foreign capital than developing *minying* enterprises as a separate category. And in Xi'an, nongovernmental enterprises have lurched between a local government determined to protect SOEs and one slowly realizing the development potential of the sector.

In the conclusion, I return to the question of technological development and national policy. In order to understand national development, we have to understand local development and local cultures. The state exists not only in central ministries, but also diffused throughout the economy. This local focus also has implications for how we understand development and economic transition in China more generally as well as in other countries. Finally I look at how concerns about technological development have been intertwined with military and strategic issues in East Asia, and especially in China.

CHAPTER TWO

Local Governments and Technological Innovation

The national policies that created nongovernmental high-technology enterprises (*minying keji qiye*) did little to define what a "nongovernmental" enterprise actually was. This ambiguity spanned the reform period and opened the door to local variation. Officials in Beijing, Shanghai, Xi'an, and Guangzhou molded national technology policy to fit local conditions. Wanting to make information industries a "pillar industry" (*zhizhu chanye*) of the local economy, Beijing focused on smaller, hybrid enterprises, Shanghai concentrated on large business groups, Xi'an fluctuated between the two, and Guangzhou paid little attention to either, content to let the market drive development.

Why did Shanghai and Beijing adopt such different methods of fostering technological innovation? Why did the city of Xi'an, blessed with a large concentration of public research institutes, lag behind in creating policies to support nongovernmental enterprises? The tools local governments adopted to foster enterprise growth—public-private cooperation, technical training, or finance to small companies—depended on the structure of the local government and the nature of its relations with social actors. Local officials operated within institutional contexts that varied along three dimensions: the distribution of authority between the central and the local government, the balance of power among local government institutions, and local state-society relations. Close links between a local economy and national development plans could reduce the space for the growth of nongovernmental enterprises. Shanghai in particular was extremely important to the national economy and so less able to experiment with new economic forms. Within governments, the strength of local science commissions mattered. If not marginalized by the planning or economic commissions, science commissions defended and advanced the interests of nongovernmental enterprises. Finally, local officials had to consider who was on the ground, and who would actually establish new enterprises. Were people with S&T talent staying in the city, or were they leaving for more vibrant economies? How close were the ties between local officials and the managers of state-owned enterprises (SOEs)? In Shanghai and Xi'an, for example, officials were both highly dependent on the revenues of SOEs and closely linked to their managers. Technology entrepreneurs had few connections to local cadres and so limited means of influencing policy.

The policies adopted by local governments were not immediately obvious given institutional constraints. Instead local cultures helped guide policymakers toward specific solutions to the problems of organizing technological development and made some types of enterprise systems more likely than others.[1] As AnnaLee Saxenian argues, "Institutions shape and are shaped by the local culture, the shared understandings and practices that unify a community and define everything from labor market behavior to attitudes toward risk-taking."[2] In China, these local cultures provided guidance to policymakers and entrepreneurs about how enterprises should be organized, how enterprises should relate to each other, and how they should interact with the local government. Local cultures shaped economic strategies by defining the problems to be addressed and by delimiting the solutions to those challenges.[3]

Many of these shared understandings were woven into local institutions by the central government, historical remainders of the process of state building, and socialist planning. Since localities had their own development needs, officials often modified these institutions to meet local conditions, producing distinct development histories. Confronting a new sector, local officials often reproduced the ideas about economic order and development that had guided them before economic reforms began; faced with new policy dilemmas, local officials designed institutions around the principles of existing institutions.[4] Shanghai in particular confronted IT development as another opportunity to use the instruments of the central plan, and so organized large business conglomerates. These practices were embedded in local institutions, but this did not mean that they could not be recast through the interaction of local officials, high-technology entrepreneurs, and central state actors. Local officials in Guangzhou, for example, promoted a policy of nongovernmental development outside of their institutional constraints and even in Shanghai the ideas used to organize enterprises have changed as policy and technology enterprises evolved.

The national policies supporting *minying* enterprises have gradually evolved from simply acknowledging the existence of these enterprises as something different from state-owned enterprises to the more recent design of venture capital funds and tax breaks specifically meant to promote nonstate sector innovation. During the late 1990s policymakers became es-

1. As Kathryn Sikkink argues, ideas help cause political and economic outcomes by giving meaning to situations and thus promoting and permitting purposeful action. *Ideas and Institutions: Developmentalism in Brazil and Argentina* (Ithaca, N.Y.: Cornell University Press, 1991), 19.

2. AnnaLee Saxenian, *Regional Advantage: Culture and Competition in Silicon Valley and Route 128* (Cambridge, Mass.: Harvard University Press, 1994), 7.

3. Frank Dobbin, *Forging Industrial Policy: The United States, Britain, and France in the Railway Age* (New York: Cambridge University Press, 1994), 20.

4. Ibid., 3.

pecially worried that China would miss the next wave of technological innovation, and so the policy environment for nongovernmental enterprises greatly improved. By 2001, the government appeared less concerned with favoring state-owned enterprises and more focused on creating technologically advanced enterprises of any type of property rights structure.

This chapter concludes with a discussion of the implications of regional economic systems for countries trying to make the transition from central planning and state socialism to more open markets and capitalism. Many of the original reform projects for transitional economies, especially in Eastern Europe, required a "big bang," a rapid dismantling of state institutions and a move to free markets as quickly as possible. The more gradual process in China and the uneven experiences of reform in Eastern Europe have forced a rethinking of this strategy. Supporters of a more incremental approach to economic transitions have argued that it is not enough that the state pulls completely out of markets as quickly as possible; rather the state must involve itself in the difficult process of creating institutions that support markets. Yet, building market-supporting institutions may be easier said then done. Even when government officials have the desire and the political support to build new institutions, they often find their range of policy options narrowed by past institutional arrangements. The future cannot be made independent of the past. The creation of nonmarket institutions in transitional economies is highly path dependent.

An examination of the Chinese case suggests the path-dependent nature of economic transition and points to two areas that should receive greater attention in path-dependence arguments. First, path dependence is both national and subnational; localities also have their own development histories. Institutional resources differ within a country as much as across nations, and so actors within the same national economy can be constrained in different ways. Second, IT development in China demonstrates that path dependence is both institutional and ideational. Path-dependence approaches have tended to concentrate on concrete institutions such as science and technology commissions or collective enterprises. In order to understand how these more identifiable organizations shape local actors, we must also understand the local cultures in which these institutions are embedded.

NATIONAL SCIENCE AND TECHNOLOGY POLICY: CREATING *MINYING* ENTERPRISES

Even the most advanced industrial countries have had difficulty creating coherent policies to foster the development of new technology firms. These countries have faltered in a number of areas, including the coordination of research projects between disparate organizations, the provision of adequate funding for technology research and development, and the commercialization of new technologies. In China, the difficulties of policy-making for technology enterprises were compounded by the simultaneous transition away from a centrally planned S&T system and a centrally planned economy. The institutional and ideological legacies of these two systems shaped the development trajectory of nongovernmental technology enterprises during the 1980s and 1990s.

The adoption of the Soviet strategy of economic development in China in the 1950s concentrated the means of production in the hands of public or state ownership. Compulsory contracts and other methods of state regulation restricted and eventually eliminated most private industry and commerce. Central planning regulated the distribution of capital goods, the allocation for investment, and the utilization of labor. The state exercised its planning function through the mandatory procurement and allocation of industrial products and the control of prices for major goods. Together these measures led to the eventual elimination of private enterprises and a state-owned system in which enterprises were divorced from production and managers had no responsibility for the results of the operation of their enterprises.[5]

The Chinese also based their S&T system on the Soviet model. Although the Chinese attained notable successes in the fields of computer, semiconductor, nuclear, and jet-propulsion technologies in the mid-1950s and early 1960s, the highly hierarchical and stratified structure of the science and technology establishment limited the scope and range of these achievements.[6] Research and development took place in diffuse locations with no coordination between them. Industrial ministries possessed their own autonomous research laboratories, and even these institutions communicated infrequently with each other because of a high degree of verticalization. Moreover, labs within state-owned enterprises operated under an administrative structure that put excessive emphasis on indicators of quantitative not technological output. Although meant to act as a supervising agency, the Chinese Academy of Sciences (CAS) did little to coordinate or direct the research undertaken by its member academies or to apply this research to industrial production.

In addition to these institutional limitations, an extremely linear model of R&D influenced central policymakers. This model (known as "technology push") viewed technological development as following naturally and easily from basic and applied research to technological development and eventually to innovation. Technology planners saw innovation as an organized collective activity, governed by research laws that were knowable and

5. Dwight Perkins, *Market Control and Planning in Communist China* (Cambridge, Mass.: Harvard University Press, 1966); and Audrey Donnithorne, *China's Economic System* (London: Allen and Unwin, 1967).

6. State Science and Technology Commission, *Guide to China's Science and Technology Policy: 1986* (Beijing: State Science and Technology Commission, 1987), 15–16.

thus amenable to planning.[7] Each of the discrete steps in the path required a supporting organization, thus reinforcing institutional redundancy.

The reforms begun in 1978 addressed the ideological and institutional barriers to economic and technological development. Most dramatically, the "Open Door Policy" attracted advanced foreign technology and management techniques in return for access to China's domestic market. The earliest investment projects were carefully chosen to raise the level of technological and managerial expertise in key areas like oil production. More generally, China's transition to a more market-oriented economy and a more commercially driven S&T system followed two parallel paths: the decentralization of fiscal and managerial control to lower-level officials and the redefinition of the system of ownership in the urban economy. Local governments, administrative departments, and a wide range of other public institutions were delegated greater authority over and responsibility for their budgets.[8] Free from no longer having to meet production quotas handed down from above, enterprises could now actively exploit market opportunities.

This decentralization was accompanied by a relaxation of the restrictions on private enterprise first within the agricultural, then the urban economy. At the beginning of urban reforms, private enterprises were expected to increase job opportunities for the unemployed and to improve the provision of scarce resources and services. Most of these enterprises were designated as *getihu,* or individual, and were restricted to seven employees from outside the employer's family and to sectors like services and handicrafts. By the mid-1980s, this program of property rights reform expanded to include allowing the creation of private (*siying*) enterprises, which were less restricted in the sectors they were allowed to enter and the number of employees they were allowed to hire, and to changes in the ownership of state-enterprises themselves. More radical reformers wanted to redefine the concept of state ownership so as to give factory managers greater freedom from government interference and more autonomy to respond to market incentives.[9]

By the mid-1980s Chinese decision-makers had become increasingly aware of the need for fundamental reforms of the S&T system. The earliest reforms had sought to repair the damage of the Cultural Revolution by establishing a national research plan, promoting international exchanges,

7. Richard Suttmeier, "Party Views of Science: The Record from the First Decade," *China Quarterly* 44 (October 1970): 146–68.

8. Christine Wong, "Fiscal Reform and Local Industrialization: The Problematic Sequencing of Reform in Post-Mao China," *Modern China* 18, 2 (April 1992): 197–227; Barry Naughton, *Growing out of the Plan: Chinese Economic Reform, 1978–1993* (New York: Cambridge University Press, 1995).

9. Harry Harding, *China's Second Revolution: Reform after Mao* (Washington, D.C.: Brookings, 1987).

and providing preferential treatment for scientists and technicians, but did little to move beyond re-creating the science establishment that had existed in the 1950s.[10] Research and development continued to be characterized by political and administrative confusion. Labor and capital remained highly immobile. In many factories, managers were motivated by political considerations, not scientific or managerial measures of effectiveness.

Even when technological innovations did occur, the R&D establishment had little incentive to introduce them into the economy. The organizational structure prevented interchanges between the different spheres of production and research, and an undefined ownership structure ensured few products moved into production. In a "socialist commodity economy" no one could be sure if an innovation was the property of the scientist, the scientist's home institution, or the state. Most enterprises treated technology as a public good and refused to reimburse the S&T unit from which the technology was transferred.

In order to overcome these impediments, policymakers applied many of the same measures first developed in the course of economic reforms to the S&T system. In March 1985, the party issued a document clarifying the reforms believed necessary to create the organizational structures capable of ensuring that the S&T system made contributions to economic growth. The "Decision of the Central Committee of the Chinese Communist Party [CCP] Concerning the Reform of the Science and Technology Management System" reformed the system of fund allocation and cut state budgets for research institutes, expanded technology markets, and encouraged cooperation between research and production units.[11]

The 1985 decision also mentioned and praised technology enterprises. Faced with increased financial pressure and reduced budgets, the decision encouraged universities and independent research institutes to take advantage of the broadening of the parameters of accepted ownership structures and launch their own commercial ventures. As long as they fulfilled their primary work tasks and did not encroach on the technical or economic interests of their original work units, S&T personnel were allowed to "engage in spare time technological work and consulting services." For such work, they were entitled to payment, but they would be required to hand over part of their income if they used technological results, data, or equipment drawn from their home units. Although the first nongovernmental enterprises actually predated the decision by five years, the decision provided increased incentives for and political justification to public research institutes to establish new enterprises.

Many of the most famous and successful *minying* enterprises originated

10. Tony Saich, *China's Science Policy in the 1980s* (Atlantic Highlands, N.J.: Humanities Press International, 1989), 22.

11. "Decision of the Central Committee of the CCP Concerning the Reform of the Science and Technology Management System" in Saich, *China's Science Policy,* 161–64.

from the Chinese Academy of Sciences during this time. With budget reductions of almost 70 percent, branches of the academy established dozens of spin-off enterprises.[12] CAS acted as an incubator for the first spin-off companies, granting permission for commercial initiatives through sometimes very ambiguous policies that dictated the relationship between the central institute and the company.[13] CAS typically advanced capital to these enterprises to be repaid once the company became commercially successful. In many instances, CAS also provided office space in its buildings.

In the context of these reforms, nongovernmental enterprises served a number of functions. They offered their sponsoring public institutions extra budgetary sources of income and could significantly raise the incomes of its employees, while relieving the main enterprise of excess personnel.[14] With their budgets reduced, more and more Beijing research organizations set up spin-off companies; by 1987 the ratio of revenue derived from budget to income rose from 1:0.37 to 1:1.61.[15] In more technologically intensive sectors, these enterprises exposed research institutes to market pressures.

In addition, and perhaps most important, Chinese leaders were increasingly aware that the revolution in information technology that was beginning in the West was a process driven by private enterprises.[16] Re-creating the institutional features of these enterprises in China was, at the time, impossible. Besides lacking the economic scale, R&D resources, managerial talent, and experience in domestic and international markets of these foreign firms, China was still not politically ready to promote and nurture private enterprises. Nongovernmental enterprises were a compromise, the best way to approximate the characteristics of innovative firms in the West and remain true to the ideological requirements of China's development.

The 863 Plan

Although the first nongovernmental enterprise emerged in 1980, other than a brief mention in the 1985 decision, these technology enterprises attracted little attention at first from central leaders. Instead, influenced by Japanese and Korean development models, central planners focused on China's ability to develop critical, or core, technologies. Analysts argued

12. Interview, B25, 30 May 1997. See also Erik Baark, "High-Technology Innovation at the Chinese Academy of Science," *Science and Public Policy* 15, 2 (April 1988): 85.

13. Zhou Peirong, "Nongovernment Technical Industry Booms," *Xinhua,* 20 November 1990, in *Foreign Broadcast Information Service-China* (hereafter FBIS-CHI), 28 November 1990.

14. For these goals in another national context, see David Bernstein, "Spin-Offs and Start-Ups in Russia: A Key Element of Industrial Restructuring," in *Privatization, Conversion, and Enterprise Reform in Russia: Selected Conference Papers,* ed. Michael McFaul and Tova Perlmutter (Stanford, Calif.: Center for International Security and Arms Control, 1994).

15. "Keji Tizhi Gaige Yiding Yaojian Xiaoyi" [Reforms of the science and technology system will certainly see benefits], *Keji Ribao* [Science and technology daily], 15 June 1987.

16. Successful enterprises in the 1980s, especially Stone, were constantly referred to as "China's IBM."

that in order to ensure economic and national security, China would have to (1) identify critical technologies; (2) adapt the technologies to local conditions; (3) diffuse new technologies to the rest of the economy; and (4) nurture domestic innovation capabilities.[17] Small, collectively owned enterprises did not seem the best way to develop these core technologies, and in March 1986 four prominent Chinese scientists wrote to Deng Xiaoping arguing for the need to promote a large-scale, state-run high-technology plan. In 1987, the National High-Technology Development Program was initiated. The plan, also known as 863 or *Baliusan* in Chinese (for the year, 1986, and month, March, it was promoted) aimed "to pool together the best technological resources in China over fifteen years to keep up with international high-technology development, bridge the gap between China and other countries in several high technologies, and, wherever possible, strive for breakthroughs."[18]

Besides targeting industries in the areas of biotechnology, new materials, lasers, energy, information, robotics, and space, 863 also introduced the concepts of peer review and a mixed method of project selection. "Expert committees" had oversight over the seven disciplines and were instrumental in the selection of projects.[19] Below the seven groups, fifteen expert task groups designated projects as deserving of support and reviewed the progress of projects on an annual basis.

From 1987 to 1989 about RMB 500 million was appropriated to 863 projects. Some ten thousand scientists and technicians from five hundred organizations participated in these projects.[20] Scientists have come from the three main sectors of the R&D establishment, with 38.1 percent from the Chinese Academy of Sciences, 26 percent from institutes of higher education, and 17.2 percent from ministerial research institutes. Successes were reported in optoelectronics, lasers, and new materials.

While the press reported a number of breakthroughs from the 863 Plan, the program was not without problems. Coordination remained difficult. Institutions competed fiercely for some of the larger projects in order to secure the resources and new equipment that came with hosting the experiments. Most important, there remained few links between research and development and production. Research institutions participating in the

17. Barry Naughton, "Technological Development, Production Networks, and IPR: How the Global Revolution in Electronics Affects China's Optimal Technology Policy," paper presented to the Sino-American Conference on Intellectual Property Rights and Economic Development, Chongqing, China, 16 September 1998.

18. The Science and Technology Leading Group of the State Council, "An Outline of the High Tech Program in China," in Richard Suttmeier, "China's High Technology: Programs, Problems, and Prospects," in U.S. Congress, Joint Economic Committee, *China's Economic Dilemma in the 1990s: The Problems of Reform, Modernization, and Interdependence*, vol. 2 (Washington, D.C.: U.S. Government Printing Office, 1991), 551.

19. "High-Tech R&D Program (Project 863) Surges Ahead," *Zhongguo Keji Luntan* [China S&T forum] 5, 18 September 1989, 8–10, in *Joint Publishing Research Services-China Science and Technology* (hereafter JPRS-CST), 4 January 1990, 1.

20. Suttmeier, "China's High Technology," 552.

programs had few official connections with enterprises, and enterprises had few incentives to look to these institutions for new innovations. Although the reforms stressed the need to link the two, many in the universities continued to be "afraid of profit."[21] Others argued that "a multitude of obstacles to transfer and extension" remained and that "the one-sided effort [to link R&D to production] by science and technology professionals has not been reciprocated by industry."[22]

The Torch Plan

The difficulties these larger, top-down projects had in commercializing new S&T products refocused policymakers on the potential role small nongovernmental enterprises could play in technological innovation. In addition to the barriers to commercialization, two other concerns motivated central policymakers. First, the central government continued to have difficulties mobilizing the large numbers of skilled scientists in the state S&T system. A 1988 survey investigating the problems facing skilled labor concluded that most talented individuals continued to languish in state-run research institutes. The solution to this problem, according to the report, was to make it easier for people to leave these institutes and start their own companies.[23]

Second, some central policymakers were growing ambivalent about relying on a labor-export strategy of development. Chinese analysts believed that low-labor costs would not last forever, and that competition in international markets was increasingly based on technology. Decision-makers gradually decided that the best solution to these two problems was not only to increase individual independence and autonomy within the state sector, but also to have more scientists actually leave the state S&T system. Decentralization of management authority and increased budget pressure were not enough to encourage most scientists to set up their own enterprises and "jump into the sea" (*xia hai*), to leave the security of their state sector job and the social welfare benefits it provided. Rather, institutions needed to be created to support scientists once they left the state system.

The central government passed a number of regulations specifically designed to encourage more individuals to leave public-sector research institutes and start their own nongovernmental enterprises, as well as to provide more extensive support for R&D and commercialization in SOEs. The most important of these plans was the Torch Plan. Initiated in May 1988, the Torch Plan broadened the sources of funds available to nongovernmental enterprises and encouraged their future growth by linking them to the de-

---

21. *Xinhua*, 10 April 1992, in FBIS-CHI, 13 April 1992.

22. "CAS Accomplishments in High-Technology Development Reviewed," *Xiandaihau*, 23 January 1990, 10–12, in JPRS-CST, 23 July 1990, 2–5.

23. "Beijing Diqu Juda Rencai Qianli Shangwei Chongfen Kaifa" [Fully develop Beijing's large skilled labor potential], *Beijing Ribao*, 3 January 1989. This point was also stressed in interview, B16, 27 April 1997.

velopment of science parks and high-technology development zones (HTDZ). Rather than having the central government arrange the delivery of commodities and allocate funds to research projects, the Torch Plan was the first large S&T plan originating from the center that was not prescriptive.[24]

Proposals selected for support under the Torch Plan had to meet a number of criteria. The technology had to be advanced and mature, conditions for batch production had to be readily available, and the results had to have good market prospects. Moreover, these products were to produce high economic benefits, with expenditure to value of output ratios of 1:5, and a production time of less than three years.[25] By 1997, the Torch Plan had approved 12,606 projects at the national and local level in five areas: new materials, microelectronics and information, energy, biotechnology, and electromagnetic devices. At the national level, 23.4 percent of all projects were "electronics and information."[26] Sixty-four percent of projects in 1997 were concentrated in eastern China.[27]

Central funding for the Torch Plan was limited, and the administering institution, the Torch High-Technology Industry Development Center (under the Industrial Science and Technology Department of the SSTC), acted like a fund-raiser and broker and not as an investor. Initial SSTC investments totaled only RMB 100 million, but by 1992 investments from all sources in the plan reached RMB 4.4 billion.[28] The plan developed a variety of funding channels, including local governments and bank lending. The Industrial and Commercial Bank was a large lender, investing RMB 200 million and $10 million, as was the Agricultural Bank of China with a RMB 400 million investment in Suzhou, Wuxi, and Changzhou.[29] Investment was also sought overseas, especially from the World Bank and the Asian Development Bank.

High- and New Technology Development Zones

In addition to broadening the funding channels for technology entrepreneurs, the Torch Plan promoted the creation of new and high-technology industrial zones to support high-technology enterprises. Through a policy of locating universities and high-technology enterprises in the same

24. Qin Shijun, "High-Technology Industrialization in China: An Analysis of the Current Status," *Asian Survey* 32, 12 (December 1992): 1129.

25. Zhang Bingfu and Tang Jun, "An Excellent Start in Implementing the Torch Plan," *Zhongguo Keji Luntan*, no. 6, in JPRS-CST, 18 November 1989, 11–14.

26. Torch Program Administrative Office, *China Torch Program Statistics for Ten Years* (Beijing: n.p., n.d.), 22–23.

27. See "Report on Torch Program during Eighth Five-Year Plan," http://www.ctp.sis.com.cn/hjj5.html, accessed April 1999.

28. Deng Shoupeng, "The Torch Plan Facing the Nineties," *Renmin Ribao* [People's daily], 19 November 1990 in JPRS-CST, 30 November 1990, 25; He Jun, "Torch Plan to Ignite," *China Daily*, 29 December 1992, 1.

29. *Xinhua*, 9 October 1992, in FBIS-CHI, 14 October 1992, 24.

area, combining research and education with production, and providing the necessary infrastructure and preferential tax and finance policies, the state expected to create environments conducive to the development of high-tech industries. These areas were to be, in the words of Song Jian, minister of the SSTC, "incubators of high and new technology."[30]

The experience of Silicon Valley, Route 128, and other science parks had a strong impact on the decision to recognize and encourage the high-technology development zones. In the early 1980s a fellow at the Institute of Physics of CAS toured Silicon Valley and other CAS researchers made similar trips during the decade.[31] A study group from CAS produced a report that was eventually reproduced in *Jingji Cankao* (Economic reference), an internal circulation report that was said to have reached General Secretary Hu Yaobang.[32] After an investigation by the Central Committee, the State Council approved the designation of the Zhongguancun area of Haidian district as the Beijing New Technology Development Zone. While Beijing was the first, Shanghai, Shenzhen, and Guangzhou soon had start-up companies and enterprise zones. By 1993, fifty-two nationally recognized zones existed throughout the country, and once designated a HTDZ, sites came under an umbrella of preferential policies and programs. The 1992 State Basic Policy for High-Tech Industrial Development Zones covered five areas of concern: taxes, finance, imports and exports, pricing, and personnel policy.[33]

Taxation. The state reduced taxes for industries within the HTDZs. During the first two years from the day they opened, new technology industries were exempted from corporate income taxes. After that time, income taxes were reduced to 15 percent for new technology industries and 10 percent for those enterprises in which 70 percent or more of the enterprise's gross output value for the year was exported. In addition, privately funded business space and attached facilities used for new technology production and operations were exempted from taxation.

Finance. Banks were to provide loans to new technology enterprises located within the zones. Enterprises oriented to the export market received preference for loans of foreign exchange. Loans within the zones were not to be divided into fixed capital and circulating capital, but rather treated as development loans. Banks within the district could withhold a certain proportion of their profits to set up loan risk funds and to set up Sino-foreign funded risk investment companies for the more uncertain ventures. More-

30. "A Guide to the Torch Plan," *Zhongguo Keji Luntan*, 6 November 1989, 7–10, in JPRS-CST, 9 January 1992, 8.

31. Ricky Tung, "The Chungkuantsun New Technology Development Zone: Mainland China's Silicon Valley," *Issues and Studies* 24, 12 (December 1988): 49.

32. See Chen Yisheng, "Zhonggou de Kexueyuan Yanjiu: 1984–1994" [Chinese science parks: 1984–1994], *Zhongguo Kuexueyuan Yuankan* [Report of the Chinese Academy of Sciences] 3 (1995): 237–40. Interview, B24, 26 May 1997.

33. This discussion on the zones draws heavily from Shao Zhengqiang, "Present Policy to Govern High, New Tech Industrial Development Zones," *Zhongguo Keji Luntan* 4 (July 1992), in JPRS-CST, 16 December 1992, 5–8.

over, relevant departments were allowed to establish venture capital funds, and the more mature zones could initiate venture capital companies.

Trade. Import-export companies were to have independent accounting and operations and be responsible for their own profits and losses. Raw materials and components imported as needed by enterprises in order to export did not require import licenses. Export products resulting from processing imported raw materials and parts were exempted from import taxes. Qualified enterprises were granted the authority to operate outside of the country and, with state permission, establish branch organizations abroad. The enterprise could retain foreign exchange generated by the export of high-tech products for the first three years.

Pricing. Prices of new products developed by high-tech enterprises not under state control (including state-set and state-guided prices) could be set by the enterprise during the trial sale period. This did not include products whose prices were set by the department in charge of pricing. The price of new products not under state control could be set by the enterprise itself.

Personnel Policy. New technology enterprises became responsible for the recruiting of their own employees, with no limits on the number of employees or the objectives of employment. Regulations exempted enterprises from taxes on enterprise bonuses and permitted them to use various systems to allocate bonuses. Technical personnel were given permission to leave the country several times a year, and their enterprises were allowed to process their visa applications.

Funding for HTDZs originated from a number of sources. With SSTC approval, RMB 1.5 billion in capital construction loan guarantees were arranged for local areas each year starting in 1991.[34] The China People's Bank arranged capital construction loans for construction of HTDZs. In most cases, municipal governments have turned to a number of sources to fund their HTDZ. In Beijing, for example, the government announced its plan to borrow $700–$800 million from foreign governments and the World Bank, $200–$300 million from foreign commercial banks, and the rest from domestic banks.[35]

Nongovernmental Enterprises and the 1993 Decision

The Torch Plan provided a needed financial and institutional boost to technology enterprises. For the first time, technology entrepreneurs could turn to a state plan for capital and to the science commission (or any other agency that managed the local HTDZ) for assistance in paying taxes, registering with commercial departments, and arranging housing permits for migrant scientists. Yet enterprises continued to face at least four barriers to

---

34. Wang Jianmin, "1.5 Billion Yuan in Capital Construction Indices to be Arranged Each Year for High and New Technology Development Zones," *Keji Ribao*, 6 March 1991, 1, in JPRS-CST, 21 October 1991, 73.

35. *China Daily*, 17 August 1992, 2.

continued growth: low managerial capacity, limited financial resources, mixed incentives for innovation, and social and economic constraints to entrepreneurship.[36] Many of the scientists employed in start-up enterprises lacked the skills necessary for more production-oriented R&D; letting scientists assume too many of the organizational tasks wasted scarce resources. Development was further restricted by the lack of capital available for most high-technology enterprises. Occasionally sums were raised through earnings from product sales and export earnings. Yet, government actors did not seem to understand that the scaling up of a production process and the "debugging" of a new technology could be the most expensive stages of development, and funds remained insufficient.

The last two limits to technology development and diffusion were more abstract and thus more difficult to alleviate. The partial nature of reforms created mixed social incentives for entrepreneurship. Although some of the first *minying* entrepreneurs became rich, many scientists feared leaving the safety and prestige of their public research institute jobs. Moreover, making money was not a characteristic admired by many Chinese scientists and intellectuals. In a speech to a meeting of a Torch study committee, Li Xue, chairman of an SSTC study group, described the problem:

> I read one of their documents on industrialization that emphasized the need to couple industrialization to education, scientific research, and the training of students. I told them not to emphasize this. Just emphasize getting rich and making money . . . you should find out which projects make money, then start work on these projects as quickly as possible. Researchers who produce scientific results are capable people, but people who are successful salesmen are also successful people.[37]

These views were reproduced in society as a whole. Many nongovernmental entrepreneurs spoke of social discrimination against *minying* enterprises; working in a SOE or state laboratory was considered a much more respectable career.

In order to address the continued financial, administrative, and social difficulties facing nongovernmental enterprises, the SSTC and the Reform Commission issued the 1993 "Decision on Several Problems Facing the Enthusiastic Promotion of Nongovernmental Technology Enterprises."[38] The decision had no specific policy impact or prescription; rather it provided a

36. Baark, "High-Technology Innovation," 85.
37. "Speech of Li Xue, Deputy Chairman, National Science and Technology Study Commission," *Keji Ribao*, 20 February 1992, 1, in JPRS-CST, 12 May 1992, 53–56.
38. Guojia Kewei he Guojia Tigaiwei [State Science and Technology Commission and State Reform Commission], "Guanyu Dali Tuidong Minying Keji Qiye Fazhan Ruogan Wenti de Jueding" [Decision concerning the problems of vigorously promoting nongovernmental technology enterprises], in *Keji Fagui Xuanbian* [Selected S&T laws and regulations] (Xi'an: Xi'an Kexue Jishu Weiyuanhui, 1996), 390–98.

six-part description of what nongovernmental enterprises were and what generally should be done to support them. Part one of the decision placed nongovernmental enterprises in a larger context, explaining that they were a desirable outcome of the reform process. According to the decision, nongovernmental enterprises had three main roles: (1) to introduce a new independent management style, based on scientific expertise and a responsibility for losses and gains, to enterprises of all types of ownership structures, including SOEs; (2) to create a new innovation system based on enterprises that were "oriented toward the market" and combined R&D, trade, and production within the same units; and (3) to slowly change an S&T system dominated by public institutes to one that embraced organizations of various ownership structures.

Part two encouraged a freer flow of S&T talent, calling on more scientists, returned students, and research institutes to set up their own nongovernmental enterprises. In part three, the decision described the need for horizontal links between science and production units, for the creation of venture capital and risk insurance funds, and for financial units to increase lending to nongovernmental enterprises. The fourth and largest section discussed the need for property rights clarification; the fifth the need to give nongovernmental enterprises a clear legal status and to start building a social welfare net for those employed in *minying* enterprises. The sixth and final section called for all levels of government, including local tax, commercial, and financial bureaus, to cooperate in coordinating *minying* development and developing policy responses to new problems as they emerged.

China and Small Enterprises

The late 1990s brought a number of changes that improved the policy environment for and focused official attention on nongovernmental enterprises.[39] Though these changes overlapped with the onset of the Asian financial crisis, the crisis itself had a relatively limited direct impact on Chinese technology policy. Its most important effect was on Chinese perceptions of Korean large business groups (*chaebol*). Many, especially Shanghai municipal officials, thought the most successful state-owned enterprises might evolve into *chaebol*. Proposals to create "enterprise groups" (*jituan*) out of the strongest state enterprises had circulated for years before the Asian crisis. The revelation of extensive problems within the *chaebol* groups during the financial crisis discredited policies to build up national champions out of SOEs. The changed perception of the *chaebol*,

---

39. The following section draws on Barry Naughton and Adam Segal, "Technology Development in the New Millennium: China in Search of a Workable Model," in *Innovation and Crisis: Asian Technology after the Millennium*, ed. William Keller and Richard Samuels (New York: Cambridge University Press, forthcoming).

however, was just one part of a broader shift in views about what kind of economy could be competitive internationally.

The dramatic explosion of the internet and related digital technologies in China received widespread coverage in the popular press, and it also influenced how Chinese leaders at the center thought about innovation. Small start-up companies appear to have been the engine of the 1990s wave of innovation in the West, and Chinese leaders were anxious not to miss out on the benefits of rapid technological change. Especially significant was the fact that many Chinese scientists and engineers were prominent in the Silicon Valley, and officials responded with policies designed to reverse China's serious brain drain and to support new business creation. Numerous localities began setting up special centers offering free rent and other benefits to lure young entrepreneurs home. Beijing, for example, announced the establishment of a Silicon Valley recruitment center in a bid to attract students to return to China.[40] In addition, China's domestic economic transformation reached a critical stage. From 1997, steady increases in production capacity combined with consistent restraint in aggregate demand led to increased competition in domestic markets. Increased competitiveness was used to shrink the public enterprise sector. Urban publicly owned industrial enterprises shed almost half of their labor force between 1992 and 1999, dropping from 81 million to 41 million total employees.[41] With these changes, the ability of the central and local governments to prop-up public enterprises or subsidize their technological expenses was severely curtailed. Moreover, China's push to join the World Trade Organization succeeded in early 2001, and WTO membership will create tough new competition for domestic Chinese enterprises.

These factors, as well as doubts about the previous policies, led to important changes in the state's relationship to nonstate enterprises. First, there was an expansion of the types of Chinese domestic enterprises deemed worthy of support. Instead of favoring large SOEs, the government now supported virtually all technologically advanced enterprises, including small, private start-ups and technology-intensive spin-offs from schools and research institutes. This reflected the important ideological changes made toward private enterprises. At the Fifteenth Party Congress in September 1997, the CCP fully acknowledged the legitimacy, contribution, and equal rights of private enterprise for the first time. In January 2000, a minister at the State Development Planning Commission announced that the government would "eliminate all restrictive and discriminatory regulations that are not friendly toward private investment."[42] And during a speech marking

---

40. "Beijing Targets High-Tech Ex-pats," *South China Morning Post*, 11 January 2000.

41. *Zhongguo Tongji Zhaiyao 2000* [China statistical abstract 2000], Beijing. This includes both state-owned industrial enterprises and urban collectives.

42. James Kynge, "Support Planned for Private Sector," *Financial Times*, 5 January 2000.

the eightieth anniversary of the founding of the CCP on 1 July 2001, Jiang Zemin announced that the party would now accept private business people as members.[43] Simultaneously, the support of nongovernmental enterprises reflected an important shift of perception. Rather than seeing private enterprises as rivals with publicly owned enterprises, these enterprises were now viewed as "national" enterprises able to compete with foreign enterprises.

Second, the nature of support changed. Government ministries were reduced in manpower and mandate; the State Science and Technology Commission was reorganized as the Ministry of Science and Technology. The government began to provide broad support for domestic enterprises designated "high-technology." This support could take the form of access to low-interest credit lines, preference in procurement decisions, or other kinds of regulatory preference or relief.

A new technology orientation toward the specific needs of smaller technology enterprises gelled in a late 1999 decision.[44] In contrast to the vagueness in the 1993 and 1995 State Council decisions, the 1999 decision called for concrete measures to foster high-tech industries and included a fund to support S&T innovation by small- and medium-sized enterprises and preferences for domestic high-tech products and equipment in government and enterprise procurements. The 1999 decision also provided a partial tax deduction for R&D expenditures; a tax exemption for all income from the transfer or development of new technologies; a preferential 6–percent value-added tax rate for software products developed and produced in China; complete VAT exemption and subsidized credit for high-tech exports; and the listing of new high-tech companies on the Shanghai and Shenzhen stock exchanges.

The decision also called for developing venture capital companies and funds. To stimulate venture capital, China changed accounting regulations on how registered capital was calculated and addressed problems of the public sale of companies (or listing on stock markets) in order to provide an exit option for initial investors. Chinese company law formerly decreed that a maximum of 20 percent of an enterprise's registered capital could be granted for the contribution of intangible "technology." Originally developed to increase the bargaining power of Chinese enterprises negotiating with technology-rich MNCs, the 20 percent cap became part of domestic company law as well.

In addition, a Law on Investment Mutual Funds was passed, and the Venture Capital Law was listed in the law-making plan of the People's Congress for the next five years. A "growth enterprise market"—like Nasdaq in the United States—that allows new high-tech companies to raise needed

43. John Pomfret, "China Allows Capitalists to Join Party," *Washington Post*, 2 July 2001.

44. The following paragraphs rely heavily on the website of the State Council Development Research Center, http://www.drcnet.com.cn, accessed March 2000.

funds has been approved in principle. As of July 2001, no such market existed on the mainland, but the government has encouraged domestic high-tech companies to list on the Hong Kong Growth Enterprise Market.[45]

BUT WHAT IS A *MINYING* ENTERPRISE EXACTLY?

Even after the 1985 decision, the implementation of the Torch Plan in 1988, and the growing media coverage of successful enterprises in Beijing, the exact nature of a nongovernmental technology enterprise remained open to debate. Western observers of these enterprises differed about which enterprises to include in this sector; Chinese reports on nongovernmental enterprises included private, collective, and state-owned enterprises.[46] Although by definition nongovernmental enterprises should be distinguished from the state sector by differences in property rights, the distinction was more usually an act of interpretation.

The problematic definitions for these enterprises came from interpretations of the meaning of the terms "high tech" and "nongovernmental." High technology, if not the easier definition to make, was at least the most clearly demarcated. Yet, the State Science and Technology Commission did not standardize even this area until a 1993 ruling.[47] Most simply, the SSTC decided that enterprises must have a specific level of technological sophistication. State regulations determined what percentage of employees must possess higher-level graduate degrees and how much should be spent on R&D for an enterprise to be called "high-tech." Regulations further decreed that high-tech firms had to produce products in one of the critical technologies listed by the SSTC: biotechnology, information industries, new materials, new energy sources, aerospace, and environmental technologies.

The meaning of "nongovernmental," or *minying,* was much more vague.[48] Enterprises came from all sectors of the economy, including the state and collective sectors. Chinese accounts of the *minying* were most clear that these enterprises were not private, but fell under a whole range

45. The 1999 "Guidelines on Examining, Approving and Supervising Applications of Domestic Enterprises for Listings on the Hong Kong Growth Enterprise Market" is supposed to apply to all types of enterprises, whether state owned or private. Beijing Yuxing InfoTech Co. Ltd., was the first company approved by the China Securities Regulatory Commission to list in 2000. Yu San, *China Economic News,* 21 February 2000.

46. Scott Kennedy, "The Stone Group: State Client or Market Pathbreaker," *China Quarterly* 152 (December 1997): 746–77.

47. Interview, B38, 24 July 1997; see also Guojia Kewei he Guojia Tigaiwei, "Guanyu Dali Tuidong Minying Keji Qiye Fazhan Ruogan Wenti de Jueding."

48. *Minban,* which preceded *minying,* can be traced back to Yenan, where it referred to local institutions that were run by and for the people. The concept embraced popular participation in and community control of areas formerly dominated by the bureaucracy. See Mark Selden, *The Yenan Way in Revolutionary China* (Cambridge, Mass.: Harvard University Press, 1971), 267.

of property rights.[49] The only thing reports seemed to agree on was that "our definition of nongovernmental is not equivalent to what they call 'private' in other countries. Our 'nongovernmental' has public ownership as the main part."[50] Even as late as 1998, this point needed reinforcement. A report in *Science and Technology Industry in China,* a journal published by the SSTC, proclaimed that the secretary of the Hubei Provincial Science Commission had recently stated that *minying* should not be confused with private property.[51] In fact, the article continued, *minying* was "clearly not a property rights concept."

Enterprises may be *minyou, minying* (owned by the people, run by the people); *guoyou, minying* (owned by the state, run by the people); or even *dayou, minying* (owned by the university, run by the people). Spin-off, privately owned, and collective enterprises, although still maintaining ties to governmental actors, most clearly fell under the heading of nongovernmental. Yet it was not unheard of for state-owned enterprises to insist that they were *minying* because they were "oriented toward the market" and their management structures and strategy were more flexible than traditional state-owned enterprises.[52] According to one source in the State Science and Technology Commission, nationally about 20 percent of all *minying* enterprises were state owned in 1997.[53]

Within the government itself, questions existed about the utility of the term. In most areas outside of the high-tech zones, *minying* did not appear as one of the possible types of property rights for tax collection and other regulations. *Minying* also did not appear in the national ownership categories for income tax rates; these categories were SOE, collective, private, and joint venture. At a meeting in Zhuhai in 1998, the Guangdong Department of Industry and Commerce demanded the end of the use of the term because it did not appear in the Fifteenth Congress report or the Constitution.[54]

Yet specialized journals and entrepreneurial organizations existed for, articles continued to be written about, and individuals identified themselves as working in *minying* enterprises. What then identified an enter-

49. Duan Ruichun, "Dui Guojia Kewei, Guojia Tigaiwei 'Guanyu Dali Fazhan Minying Kejixing Qiye Ruogan Wenti de Jueding' de Shuoming" [An explanation of the SSTC and Reform Commission's decision on promoting nongovernmental technology enterprises], *Zhongguo Minban Keji Shiye* 62 (July 1993): 40–41.

50. "Zhongguo Minying Keji Qiye Fazhan Zhanlue yu Zhengce Yanjiu Taohui" [Discussion of Chinese nongovernmental enterprises development strategy and policy research], *Zhongguo Minban Keji Shiye* 11 (1995): 13.

51. "Gu Zhijie Shuo 'Minying' Butong 'Siyou'" [Gu Zhijie says that "nongovernmental" is not the same as "private"], *Zhongguo Keji Chanye Yuekan* [Chinese technology industry monthly] 4 (1998): 30.

52. Interview, S15, 23 April 1996.

53. Interview, B38, 24 July 1997.

54. "Gongshang Bumen Yaoqiu Wucheng 'Min Ying'" [Industrial and commercial bureau does not want to use "nongovernmental"], *Zhongguo Keji Chanye Yuekan* 4 (1998): 30.

prise as nonstate? According to published reports and interviewee comments, even SOEs were nongovernmental if they were run by the principles of a *minying* enterprise. The problem with an enterprise like Great Wall, a state-owned computer manufacturer, was not that it was state owned; the problem with Great Wall was that it was managed like a state-owned enterprise.

For the purposes of this book, *minying* enterprises were enterprises staffed by individuals who understood technological development and were free from outside interference. These principles were summarized in two slogans frequently repeated in publications about nongovernmental enterprises: the "two no's" (*liang bu*), and the "four self principles" (*si zi*). *Minying* enterprises did not want be part of the plan and did not want government interference in managing personnel. Moreover, they were responsible for their own development, organization, financing, and all profits and losses.[55]

In short, state and collective enterprises had clear lines of subordination; they had administrative relations (*lishu guanxi*) with their supervisory agencies (*zhuguan bumen*). Members of the supervisory agencies sat on the board of directors and had a say on development, research, personnel, and salary decisions. *Minying* enterprises either did not have supervisory agencies or did not have tight administrative ties to the agencies to which they were linked. That is, *minying* enterprises were not linked to bureaucratic superiors who had the presumptive right to intervene in the enterprise's business operations.

The central government encouraged and took advantage of the ambiguity in the definition of these terms. Decision-makers hesitated to expose the high-tech sector to coordinated reform measures. As the political scientist Peter Cheung argues, "The incrementalist approach to reform enables provincial leaders to interpret central policies more liberally than if central policies are more clearly defined and their implementation more effectively monitored."[56] At a 1997 meeting in Guangzhou, a representative from the SSTC, in response to pressure for a clear definition of what *minying* actually meant from a local official, answered: "If we clarify what it means to be *minying,* we are going to have problems. You keep doing what you want, and then we can see what works best."[57] Ambiguity allowed the central government to proceed simultaneously with various, and at times, widely divergent forms of property rights in the reform of the S&T system.

---

55. The *si zi* are: *ziyuan zuhe, zizhu jingying, zize yingkui,* and *ziwo fazhan.* See Wang Fengyun, "Yinru Minying Jizhi Fahui Qiye Jishu Kaifa Jigou Zuoyong" [Draw in the nongovernmental system, unleash the ability of enterprises to develop technology], *Keji Ribao,* 29 July 1996, 8.

56. Peter Cheung, "Introduction: Provincial Leadership and Economic Reform in Post-Mao China," in *Provincial Strategies of Economic Reform in Post-Mao China,* ed. Peter Cheung, Jae Ho Chung, and Zhimin Lin (Armonk, N.Y.: M. E. Sharpe, 1998), 14.

57. Interview, G15, 6 July 1998.

This may have been the best strategy to take with the information industries; given the lack of any obvious formula for intervening in the sector, it allowed Chinese leaders to see which patterns of support worked best.

THE INSTITUTIONAL CONTEXT OF LOCAL DEVELOPMENT

The macrolevel policies described above created the political and organizational space for nongovernmental enterprises. Ambiguity in the definition of *minying* and an incremental approach to reform meant it was left to local governments to decide how they were going to populate that space. The decisions officials made were heavily influenced by the local political and institutional context, which consisted of three factors: the distribution of authority between the central and local government, the balance of power among local government institutions, and the nature of local state-society relations.

The institutional presence and power of the central government in the locality helped determine how and when nongovernmental enterprises were supported. Although national technology policy did not recognize provincial differences, the creation of technology enterprises was increasingly linked to a larger economic strategy that eventually encouraged regional differences. Maoist development strategy, for political and strategic reasons, attempted to eliminate regional differences in growth, or at least prevent the urban-rural gap from widening. To reduce these differences, the central government redistributed revenues collected in highly developed areas like Shanghai to poorer interior provinces. In addition, scientists and other skilled technicians were frequently "sent down" to rural areas to work on development projects.

The reforms started under Deng recognized and exploited regional comparative advantage. Chinese leaders publicly declared that some areas were going "to get rich first." The Coastal Development Strategy opened Guangdong and other southern provinces to investment from Hong Kong and Taiwan, and the 1985 strategy for Shanghai focused on developing the city as an international commercial, financial, and service center. In 1992, interior areas like Xi'an were allowed to offer preferential tax policies in order to attract foreign investment. And in 2000 the center launched a "Go West" campaign aimed at accelerating the development of western China.

These development strategies reflected a growing consensus among central and local leaders about the role specific regional economies might play in the national economy. While policies designed for nongovernmental technology enterprises were to be implemented uniformly across the country, economic policy was not. Decentralization was not a uniform policy, and in fact deliberately slighted some areas rather than others. As the political scientist Dorothy Solinger argues, the varied success of different regions depended quite heavily on how, when, and how often the central government allocated resources.[58]

Shanghai, for example, occupied a unique position in the Chinese economy under reforms, and its financial burden was higher than any other locality in the country. The size and importance of its remittances to the center significantly reduced Shanghai's space for institutional innovation. Moreover, as Lynn White argues, state planning had a wide and pervasive influence on the city. Factory managers faced more rules and regulations, and state agencies were heavily involved in organizing product flows. Shanghai gained a reputation, according to White, of being a city that followed the central government's rules.[59] Guangdong's more autonomous relationship with the center and its geographical proximity to Hong Kong made the adoption of more private forms of organization and a heavy reliance on foreign investment more likely. In addition, policy shifts at the center opened and closed the political space for local experimentation in the organization of technological development. For example, Deng Xiaoping's southern inspection tour in 1992 signaled continued support for reform and initiated new stages of *minying* enterprise growth throughout the country, especially in Shanghai.

The organization of power within local governments was also a factor in deciding how technology policy was implemented. In every locality, officials within the science and technology commission were more sympathetic to and had closer links with *minying* entrepreneurs than officials in the planning and economic commissions. In cities like Shanghai where the planning commission tended to dominate, nongovernmental entrepreneurs had a more difficult time in getting the ear of local authorities. These different commissions also competed over the creation and management of high-technology parks.

Finally, the nature of state-society relations at the local level, especially the relationship between state and S&T personnel, influenced which development strategies were adopted. Some local governments were more limited in the range and type of partners they could choose from for technological development. Until 1978, Shanghai suffered from an outflow of both private entrepreneurs and skilled technicians. In Xi'an, a large number of scientists and S&T personnel worked in military R&D institutes, and these individuals had different attitudes toward starting their own enterprises than those employed by independent research institutes. Guangzhou especially lacked the quantity and quality of S&T institutes present in other parts of the country. In 1994, Shanghai had 251 total re-

58. Dorothy Solinger, "Despite Decentralization: Disadvantages, Dependence, and Ongoing Central Power in the Inland: The Case of Wuhan," *China Quarterly* 145 (March 1996): 33.

59. Lynn T. White, *Shanghai Shanghaied? Uneven Taxes in Reform China* (Hong Kong: University of Hong Kong, 1989), 29.

search and development organizations compared to Beijing's 422; the entire provinces of Guangdong and Shaanxi, of which Guangzhou and Xi'an
are the capitals, had 289 and 191 research organizations respectively.[60] Beijing had over eighty thousand scientists and engineers, almost eight times
the amount of Guangdong, three times Shaanxi.[61]

Also entrepreneurs relied on local social resources to build their enterprises. In Beijing and Xi'an, local officials and technology entrepreneurs
relied on well-developed school networks, while civil organizations were
more inchoate and diffuse in Guangzhou. Finally, development strategies
were shaped by the relationship among state personnel, especially among
SOE managers and local officials. The large concentration of SOE managers in Shanghai and Xi'an made it easier for them to lobby against the
local government supporting nongovernmental enterprises.

DEVELOPMENT AND LOCAL CULTURE

The interactions of local governments and entrepreneurs created regionally distinct patterns of development. These patterns of development
emerged within institutional constraints. But structural constraints alone
did not determine what types of policies were adopted and what types of
enterprises were created in each locality. As Richard Locke notes, "Strategic choices are shaped as much by the qualitative features of the socio-po-
litical context in which they are embedded as by their own organizational
resources and capabilities."[62] Institutions and local actors operate within
and are shaped by distinct local cultures. Unsure of how to develop new
sectors and how rapidly changing technologies would develop in the future, local officials, often implicitly, reverted to shared understandings and
routine practices of development to define problems, simplify complex
causal chains, and build a consensus to act on these problems.

At the most basic level, these shared understandings addressed who
should be involved in innovation, how they should be funded, and how
their S&T products should be commercialized. Officials in different regions could believe that innovation was more likely to emerge from a group
of scientists working together in a large lab, or they could look to support
the lone individual touched by moments of greatness. Once products were
ready, local governments differed on what type of organizations were re-

60. Xi'an is a mixed case. It is often said to have the third most developed S&T resources in
the country. This is discussed further in chapter 5. *Zhongguo Kexue Jishu Zhengce Zhinan* [Guide to
China's science and technology policy] (Beijing: Kexue Jishu Wenzhang Chubanshe, 1995),
420.

61. The rest of the country suffers from neglect. Nine out of thirty provinces account for
70.4 percent of total expenditure on science and technology. See *Xinhua*, 29 July 1995, in FBIS-
CHI 39, 30 July 1995.

62. Richard Locke, "The Composite Economy: Local Politics and Industrial Change in Con-
temporary Italy," *Economy and Society* 25, 4 (November 1996): 501.

sponsible for bringing these innovations to the market, fixing on university extension projects, venture capitalists, or state agencies.

Shared beliefs about the organization of technology industries also clarified how authority should be deployed in support of innovation and, perhaps more importantly, why. In effect, local Chinese technology policy revealed the broader social purposes for which local state power was to be employed and how this power related to economic order.[63] Willing to take a less interventionist role in the economy, the Beijing local government shared the authority to shape technology markets with scientists and entrepreneurs. By contrast, the Shanghai municipal government sought to retain its ability to guide development and concentrated power in large state-owned units. In Guangzhou, the local government has allowed nonstate actors to operate more freely in new markets than anywhere else. And in Xi'an, the local government has yet to decide if it should work with scientists and entrepreneurs, or if the state should remain the main engine of technological development.

These interpretations can be discovered in a wide range of planning documents, reports, and journal articles. Planning documents in Beijing stressed the role of specific individuals and enterprises, not the government. Local officials talked about supporting and further developing a "technological network" (*keji wangluo*) linking universities, enterprises, and the local government. Local government officials like Vice Mayor Hu Zhaoguang and Lu Yudeng, director of the city's Science and Technology Commission, supported the "Beijing model." The research institute of the Beijing Experimental Zone also widely publicized and promoted technological networks as the basis of Beijing's development. In contrast, Shanghai relied on many of the tools of central planning to encourage development. The city government played the dominant role as both producer and consumer of technological products. Publications from the research arm of Caohejing, Shanghai's first high-technology park, barely mentioned *minying* enterprises, stressing the need to merge state-owned enterprises into *jituan,* or *chaebol-*like conglomerates. White papers produced by the city government and the local science commission concentrated on the role of *minying* enterprises as complements to the dominant state sector.

These shared beliefs about technological development were both causal and purposive; they justified value commitments and reflected ideas about the relationship between political power and technical knowledge. In newspaper accounts about the successful development of high-tech enterprises in the Chinese press, *rencai,* or talented individuals, stand at the center of the narrative. Still it remained up to local governments to decide how to best take advantage of their *rencai.* Cadres could choose either to encourage talented individuals to set up their own independent enterprises or to redistribute management authority to S&T personnel within state enter-

63. See in a different national context, Dobbin, *Forging Industrial Policy,* 2.

prises. The scientists themselves naturally argued that their skills were key to an enterprise's success. These entrepreneurs deployed *rencai* not only as a justification of why local governments should support nongovernmental enterprises but also of how rewards should be distributed within an enterprise. Opposed to measures based on seniority or political reliability ("redness"), S&T personnel argued that *rencai* was the only legitimate measure by which to distribute salaries. Moreover, they increasingly lobbied for local governments to acknowledge the value of their *rencai* by approving plans to convert technological know-how into stock holdings within companies. Debates over *rencai* reflected distinct views about the creation and role of enterprises in new markets, the local government's relationship with these enterprises, and authority relations within enterprises.

CHINA AND TRANSITIONAL ECONOMIES

The importance of inherited industrial and institutional structures has a resonance outside the Chinese case, especially for the transitional economies of Eastern Europe. The debate in the literature on transitional economies has often been framed as the choice between rapid change, or shock therapy, and more gradual reform. Economists have long argued that the reform of a planned economy requires that the state remove itself from economic activity as quickly as possible. Advocates of shock therapy like Jeffrey Sachs argue for rapid market reform simultaneously across as many sectors as possible. Macroeconomic stabilization and privatization of state-owned enterprises should accompany price and trade liberalization.[64] All of these reforms must be enacted quickly; moving more gradually risks creating nonfinancial incentives for investment or opportunities for rent-seeking behavior.

In contrast, the Chinese case suggests a more incremental approach to reform. Compared to Russia and Eastern Europe, China took a more experimental tack, frequently reversing policy decisions and then trying new things. Policymakers attacked one set of problems but left others alone; price controls were lifted, but soft loans for SOEs remained. Lacking an overall blueprint for reform, policymakers were free to move on to new areas when one proved particularly obstinate.[65] China, in the words of John McMillan and Barry Naughton, "muddled through."[66] This muddling through had at least two clear advantages. First, it allowed policymakers to

64. See for example, Jeffrey Sachs and Wing Thye Woo, *Understanding China's Economic Performance* (Cambridge, Mass.: Harvard Institute For International Development, 1997), 5.

65. Gary Jefferson and Thomas Rawski, "Enterprise Reform in China Industry," *Journal of Economic Perspectives* 8, 2 (spring 1994): 47–70; and Dwight Perkins, "Completing China's Move to the Market," *Journal of Economic Perspectives* 8, 2 (spring 1994): 23–46.

66. John McMillan and Barry Naughton, "How to Reform a Planned Economy: Lessons from China," *Oxford Review of Economic Policy* 8, 1 (spring 1992): 131.

pick the best out of a range of policy options. Successful local experiments were adopted nationally, failures dropped by the wayside. Second, a more gradual process allowed both the expansion of the market and the creation of the institutions that support market activity.

For our purposes, what is interesting about the debate between the gradualists and shock therapists is less the policy prescriptions they suggest and more about the different views of growth and of markets they hold. Inherent in the more incremental approach to reform is an appreciation that the state plays an important role in the construction of market institutions and a view of markets not as an economic abstraction but rather as institutions that require rules and widely shared customs and practices. This view of economic transition recognizes that the state cannot rapidly or completely withdraw from the market because it is intrinsically involved in defining property rights, creating transparent governance structures, and standardizing management practices without which markets could not operate efficiently.

Moreover, in contrast to a more rapid vision of reform, the gradualist approach to economic transition is much more path dependent; it argues that the institutional structures of the past place constraints on the possibilities of the future.[67] The difficulty for policymakers in China or Eastern Europe was not only that markets needed to be created; it was also that they were being built on the foundations of state socialism. "Transitional economies," as Edward Steinfeld argues, "are not developing market economies; they are economies trying to develop markets."[68] These economies are already populated by actors responding to incentives from both the emerging and the past system.

Under these conditions, the range of policy options narrows considerably. As the sociologist David Stark argues about transition in Eastern Europe, "Actors who seek to move in new directions find their choices are constrained by the existing set of institutional resources. Institutions limit the field of action, they preclude some directions, they constrain certain choices."[69] Stark's point is as true for the Chinese case and can be expanded in two ways. First, the existing set of institutional resources is likely to vary within a national economy and so local actors are constrained in their own local ways. How local officials respond to national policy will be heavily influenced by local institutional structures. This has real implications for development; the institutional endowments of localities deter-

67. Stephen Krasner, "Approaches to the State: Alternative Conceptions and Historical Dynamics," *Comparative Politics* 16, 2 (January 1984): 237.

68. Edward Steinfeld, *Forging Reform in China: The Fate of State-Owned Industry* (New York: Cambridge University Press, 1998), 253.

69. David Stark, "Path Dependence and Privatization Strategy in East Central Europe," in *Comparative National Development: Society and Economy in the New Global Order,* ed. Douglas Kincaid (Chapel Hill: University of North Carolina Press, 1994), 169–98.

mine how officials are able to respond to particular development challenges and to which industries they will be more or less successful in supporting.[70]

Second, the constraints on policymakers are both material and ideational. Path dependence approaches have tended to focus on the more identifiable institutions of development like state-owned enterprises or local planning commissions and the economic incentives they create. But institutions do more than provide services or reduce transaction costs; they are also a representation of the symbolic and political value of ideas, and these ideas are often reproduced as local governments address new development problems.[71] The sociologist Neil Fligstein argues, "As new industries emerge or old ones are transformed, new rules are made in the context of old rules."[72] As local governments build new institutions, historical ideas about how to organize industries continue to inform and structure policy choices across industrial sectors.

CONCLUSION

Understanding the emerging regional patterns of technology development in the Chinese national economy requires understanding what local governments did and why they did it. In particular, local governments engaged in the process of enterprise creation. Policy decisions in the realms of property rights, funding, and regulation directly impacted the internal organization of enterprises, deciding which types of enterprises were more likely to be able to respond to market opportunities. In addition, government actions influenced the shape of the emerging relations between enterprises.

The critique of explanations that rely on both institutions and culture to predict economic outcomes is that it is too difficult to disentangle the causal power of the two types of variables. Institutional arrangements and political constraints provide a more parsimonious explanation for economic outcomes. In the four cases under discussion, it is in fact difficult to separate the two threads of cultural and material variables. During the early stages of reform, government actors in Shanghai, Beijing, Guangzhou, and Xi'an did not adopt development strategies that contradicted their political or institutional endowments.

Two points can be made, however, about why the cultural content of local development remains important. First, changes across time and

70. Adam Segal and Eric Thun, "Thinking Locally, Acting Globally: Local Governments, Industrial Sectors, and Development in China," *Politics & Society* 29, 4 (December 2001): 557–88.

71. Emmanuel Adler, *The Power of Ideology: The Quest for Technological Autonomy in Argentina and Brazil* (Berkeley: University of California Press, 1987), 15.

72. Neil Fligstein, "Markets as Politics: A Political-Cultural Approach to Market Institutions," *American Sociological Review* 61 (August 1996): 661.

within cases provide an opportunity to control for structural changes. In cases where structural variables have changed but approaches to nongovernmental enterprises remained similar, it is easier to see the independent influence of local development practices. In Shanghai, for example, the political environment changed. Until the early 1990s, local leaders adopted a conservative, obstructionist attitude towards nongovernmental enterprises. The main engine for local development was to be the large enterprise group. This reflected the role SOEs played in the local economy, the role Shanghai played in the national economy, and the center's unwillingness to allow Shanghai to adopt a more aggressive reform strategy. After Deng's inspection tour in 1992, Shanghai was allowed and encouraged by the center to adopt more market-oriented reforms, and the number of nongovernmental enterprises exploded from 1,826 in 1992 to 5,644 in 1993.[73]

The important point here is not only the quantity of enterprises, but also the quality of the Shanghai municipal government's interactions with the enterprises.[74] The number of enterprises certainly increased, but these enterprises remained at the margins of the local economy, isolated from each other and from government actors. Some district officials began paying more attention to these enterprises, adopting Beijing-style financial and supervisory policies. But the municipal government demonstrated a continued ambivalence toward small enterprises; entrepreneurs still complained they could not secure loans or clarify property rights. The reforms that accompanied Deng's trip may account for the increase in the quantity of nongovernmental enterprises, but the continuity of a development strategy based on government intervention and large enterprises explains why the ever-growing number of *minying* enterprises still faced difficulties in developing technological capabilities.

Second, local development cultures can be viewed as complementing not replacing material explanations. Twenty years into the reforms it is easy to look back and say that it was likely that Shanghai, given its place in the national economy, would view nongovernmental enterprises as marginal to the state-owned economy. But at the time, local officials had to react to their material constraints, they had to construct narratives of development, and this was a political process. In Peter Hall's words, "Policy making is based on creation as well as constraint," and in order to trace the process of creation we need to focus on the cultural component of local economic systems.[75] Given several possible development paths to follow, the problems local officials chose to tackle and the types of solutions they believed likely to succeed were heavily influenced by the local culture. Institutional

73. *Shanghai Keji Tongji Nianjian: 1995* [Shanghai S&T yearbook: 1995] (Shanghai: Shanghai Tongjibu Chubanshe, 1996), 38.

74. This is discussed further in chapter 4.

75. Peter Hall, "Conclusion: The Politics of Keynesian Ideas," in *The Political Power of Economic Ideas,* ed. Peter Hall (Princeton, N.J.: Princeton University Press, 1989), 362.

arrangements provided the broad outlines of what might be possible. Shared beliefs about how development should be organized linked specific policies to economic outcomes.

In the remainder of this study, I analyze how the local political economy in Beijing, Shanghai, Guangzhou, and Xi'an shaped technology enterprise development. In each case, I develop a local history of *minying* enterprises using planning documents, interviews with high-tech entrepreneurs and local government officials, and case studies of selected enterprises. These histories highlight local government actions, focusing on how local officials defined property rights, funded high technologies, and regulated high-tech enterprises. In Beijing, the local government institutionalized an enterprise system centered on small, autonomous enterprises. Shanghai local officials did not want to displace SOEs from the commanding heights of the local economy. Guangzhou and Xi'an have mixed their strategies, trying to balance the needs of nongovernmental enterprises with the demands of the rest of the local economy. In addition, by analyzing the relationship between local and central governments, the institutional configuration of the local state, and emerging state-society relations, I explain why development strategies more supportive of nongovernmental enterprises were adopted in Beijing but not in other parts of the country.

CHAPTER THREE

Beijing: Creating China's Silicon Valley

Like other local governments in China, Beijing started the 1980s by reforming agriculture and by addressing the problems of large state-owned enterprises (SOEs) in more traditional industrial sectors.[1] After initiating these reforms, the local government then turned to the science and technology system. In particular, the local government hoped to link scientific research more closely with industrial production. Nongovernmental enterprises were meant to be a step away from the economic and institutional weaknesses of SOEs and a step closer toward the small, flexible, and privately owned technology firms found in the West, especially in Silicon Valley.

Geographically and politically a long way from Silicon Valley and its model of flexible, innovative enterprises, local government officials approached nongovernmental enterprises as the closest to that model they were going to get. Even though state-owned enterprises made up almost 20 percent of *minying* enterprises in Beijing until the mid-1990s, local officials acknowledged individual scientists and entrepreneurs, not state-owned units, as the driving force behind nongovernmental development. Policy documents and research studies produced by the local government stressed the nongovernmental nature of technological development in the city.

Government action institutionalized and reproduced a conception of technological development centered on independent and competitive *minying* enterprises. This focus on the nongovernmental expressed itself in three policy areas: (1) the funding of high-tech research; (2) types of property rights recognized by local governments; and (3) the scope and character of government supervision. In general, local authorities in Beijing were more likely to encourage multiple property rights and to direct funding to the general science budget and broad-based technologies. They were also relatively less likely to intervene in the internal management of enterprises. The manner in which these policies were implemented addressed the main needs of technology enterprises—capital, technology, and skilled personnel—while preserving management autonomy.

These policies helped create enterprises that, while not looking exactly like technology firms in the West, did not resemble traditional state-owned

---

1. Duan Bingren, "Jing Cheng Shinian zhi Lu" [The capital's ten-year journey], *Beijing Ribao*, 27 February (part 1), 28 February (part 2), and 6 March (part 3) 1989.

enterprises either. Enterprises in Beijing tended to be small and non-governmental; authority within the enterprise was based on technological accomplishments, not seniority or political reliability; and barriers between marketing, research, and sales departments were low. By 2001, nongovernmental enterprises were something in between what was hoped they would become—modern firms—and from what they emerged—state-owned enterprises.

The first scientists to set up their own enterprises did not rely solely on the local government to shape the sector. Instead, they created new institutions to secure the resources essential to their success. In the late 1980s entrepreneurs first recognized changes in the market and began discussing what both the local government and individual enterprises needed to do to react to new market forces. Eventually some of the policies and programs adopted by the local government in response to market change originated from these discussions. Much of Beijing's success was a result of the growing interaction between independent nongovernmental entrepreneurs and local officials.

Beijing's success has so far been a relative success. In 1998, Silicon Valley had 8,000 enterprises with annual turnover of $200 billion. Beijing, by contrast, had 4,000 companies that registered $5.4 billion in annual volume of business in 1999.[2] The IT sector in Beijing and in China more generally continues to face real challenges; property rights remain undefined, venture capital scarce, and governance structures underinstitutionalized. Moreover, China's entry into the World Trade Organization will open the domestic market to severe competition from much larger, more technologically advanced foreign IT firms. The local government has not been perfect in its support of the local sector; local officials have overstepped, interfering in specific enterprises, and dragged their feet, not responding quickly enough to changing market structures, especially in helping enterprises resolve ownership questions. Still, the Beijing local government has been the most successful in combining governmental support and guidance to entrepreneurs, while still allowing their efforts to shape the sector.

THE INSTITUTIONAL, ORGANIZATIONAL, AND SOCIAL STRUCTURE OF BEIJING

Compared to the other three cases discussed in this book, Beijing was uniquely blessed in both the organizational structures and factor endowments needed to promote high-technology enterprises. Before 1978, Beijing's role in the national economy was determined by the central plan. Focused on heavy industry and self-reliance, these plans sacrificed Beijing's

---

2. Antoaneta Bezlova, "China's Last Chance to Catch the High-Tech Train," online at http://www.atimes.com/China/China.html, accessed 4 December 1999.

comparative advantage to larger political concerns. After the founding of
the People's Republic in 1949, planners wanted to rebuild Beijing as the
political, cultural, and economic capital of the country with an emphasis
on creating a science and technology center with a strong industrial base.
The 1956 Urbanization Development Plan, for example, stressed the need
to balance industrial development with the city's role as political and cultural center.

This balance was lost however during the Great Leap Forward, and soon
after planners concentrated on turning Beijing into a comprehensive industrial base. Between 1957 and 1978 investment in heavy industries increased 474 percent with a particular focus on developing iron and steel.
City and central planners also concentrated on increasing the quantity of
large- and medium-sized enterprises in the city; the number of SOEs grew
from 140 in 1950, to 1,124 in 1983.[3]

With the start of the reform process, central planners began to recon-
sider their earlier emphasis on heavy industry for Beijing. In April 1980,
the State Council pointed out that "development of the Beijing urban and
rural economy should be subordinate to and serve the requirements of Beijing as the political and cultural center of the country."[4] New guidelines
emphasized the improvement of Beijing's social and environmental conditions, the development of the city's educational institutes, and the promotion of economic prosperity based on the service sector and new technologies.[5] Local officials began closing down some of the largest heavy industry
enterprises, and between 1980 and 1984 twenty enterprises ceased operation; from 1978 to 1983 the ratio of light to heavy industry increased from
34:56 to 45:55.[6] In January 2000, the city announced plans to move another 134 industrial enterprises out of downtown.[7]

The declining role of heavy industry in Beijing had at least three positive
impacts on nongovernmental growth in the city. First, unlike in Shanghai
and Xi'an, the central government was not especially worried about the
possible negative impact that experimenting with local enterprise systems
would have on the national economy. Beijing especially was never as constrained in terms of revenue remittance or fiscal burdens as Shanghai. For
example, from 1988 until 1993, a high percentage of the tax revenue collected by the central, local, and district government in Beijing was reinvested in the Beijing Experimental Technology Zone. How tax revenues

3. Han Guang, "Beijing Municipal Government's Management of State-Owned Enterprises,"
in *China's Economic Reform: Administering the Introduction of Market Mechanisms,* ed. George Totten
and Zhao Shulian (Boulder, Colo.: Westview, 1992), 77.
4. Quoted in Han, "Beijing Municipal Government's Management," 78.
5. Zixuan Zhu and Reginald Kwok, "Beijing: The Expression of National Political Ideology,"
in *Culture and the City in East Asia,* ed. Won Bae Kim (New York: Oxford University Press, 1997),
136.
6. Han, "Beijing Municipal Government's Management," 80.
7. "Beijing Drafts Blueprint for Industrial Development," *Xinhua,* 28 January 2000. This is
probably linked to the desire to improve the city's air quality for the Olympics.

were actually distributed is unclear. According to one interview, until the 1994 tax reforms 45 percent of tax revenue stayed in the zone, the remaining 55 percent was handed over to the central government.[8] After the reforms, taxes were officially divided equally between the Beijing city, Haidian district, and central governments. Yet another official argued that a new agreement made behind closed doors ensured that the distribution stayed pretty much the same.[9] In any case, given that tax remittances for the zone totaled RMB 1 billion in 1996 compared to RMB 150 million in 1988, the local and central governments were sharing an increasing bounty.[10] This situation does not approach the severe financial constraints that affected Shanghai, which remitted about 87 percent of its revenue to the center between 1949 and 1983.[11]

Second, the absence of a large concentration of SOEs also meant that the tools of industrial central planning were not as well developed at the local level in Beijing as they were in other cities, especially Shanghai. Since they were not state enterprises, Beijing did not have administrative ties to or control over many of the most important players in technological development like the local government did in Shanghai. The most active bureaucracies in *minying* development were the State Science and Technology Commission (SSTC) and the Beijing Science and Technology Commission—agencies that only had supervisory and regulatory powers. *Minying* enterprises had few if any links to the local government; almost all of the first enterprises split off from the Chinese Academy of Sciences (CAS). They drew their personnel, technology, equipment, and investment capital from the academy. Local government officials had little control over these central government agencies. In this context, the local government was limited to developing the city's social and educational infrastructure and providing social and political services to *minying* enterprises.

Finally, SOEs in Beijing were contracting, not expanding into new sectors. The earliest nongovernmental enterprises entered market areas—software, bioengineering, and new materials—that were completely new and so lacked bureaucratic organs with sectoral authority. When nongovernmental enterprises tried to enter market niches where ministries already had a strong presence, like photomechanical electronic integration,

8. Interview, B15, 24 April 1997.

9. According to this official, although the reforms were supposed to ensure that each of the three received an equal percentage of revenues, the actual distribution was 50–50 between the Beijing and central governments. The Beijing government then distributed an unknown percentage of its share of the funds to the Haidian district government. Interview, B17, 5 May 1997.

10. "Beijing Zhongguancun High Tech Zone," *Jisuanji Shijie* [Computer world] 44 (14 November 1990): 5, in *Joint Publication Research Services-China Science and Technology* (hereafter JPRS-CST), 12 January 1991, 119.

11. J. Bruce Jacobs and Lijian Hong, "Shanghai and the Lower Yangzi Valley," in *China Deconstructs: Politics, Trade, and Regionalism*, ed. David S. G. Goodman and Gerald Segal (New York: Routledge, 1994).

they faced strong competition from enterprises under the control of the Ministry of Mechanical Industry and Ministry of Electronic Industry.[12]

Social Structures and Entrepreneurship

Beijing was also extremely blessed in the large concentration of possible partners to high-technology development located in the city. The central government began concentrating S&T resources, including the most talented scientists and engineers, in the city immediately after 1949. Many of these scientists worked together for years on the same projects; some of the first products commercialized by Beijing *minying* enterprises had been under development in local universities since the 1950s. Long years of working together under difficult circumstances created a network of scientists that facilitated an esprit de corps and a sense of identity. Entrepreneurs in Beijing saw themselves as part of an intellectual tradition of questioning the old order and experimenting with new forms of organization that dated back at least to the May Fourth movement of 1919.[13] The validity of these comparisons is less important than the ease and extent to which entrepreneurs could invoke these shared memories to provide a sense of identity.

Local officials believed that the earliest stages of high-technology development were successful because Beijing was located close to but also removed from the central government. According to this explanation, local officials and entrepreneurs operated without attracting the attention of the center. As one local official put it, "It is always darkest directly underneath the lamp."[14] This explanation is partly correct, but ties to the center were an important resource that nongovernmental entrepreneurs lacked in other localities. Once encouraged to set up enterprises, scientists and university professors relied on extensive social networks to raise capital or mediate conflict with local bureaucrats. School networks, especially at Beijing and Qinghua universities, were well developed and could be used to lobby local and central policymakers for political protection. Wan Runnan, the founder of Stone, not only attended Qinghua, but was also married twice to the daughters of important leaders; first to the daughter of Liu Shaoqi, then to the daughter of Li Chang, a Qinghua graduate and member of the Central Committee.[15] *Minying* entrepreneurs relied on protection from ideological attacks from reformist segments, like Wu Mingyu in the SSTC and

12. Interview, B36, 18 July 1997.

13. In one of his more expansive moments, a consultant to nongovernmental enterprises declared that Beijing entrepreneurs were "less like Bill Gates, and more like Thomas Jefferson. Gates is interested in money. Entrepreneurs come from a scholarly tradition and ask questions that are about more than business." Interview, B29, 28 June 1997.

14. In Chinese, *"deng xia hei."* Interview, B17, 5 May 1997.

15. Cheng Li, "University Networks and the Rise of Qinghua Graduates in China's Leadership," *Australian Journal of Chinese Affairs* 32 (July 1994): 27.

officials in CAS. Moreover, political connections gave Beijing entrepreneurs a sense of how resolute the central government was in its support of nongovernmental enterprises. Given China's turbulent history of political campaigns, it was reasonable for entrepreneurs to worry that nongovernmental enterprises even with official sanction might come under attack in the future as ideological winds changed. Closer to the corridors of power, entrepreneurs in Beijing were better able to read these winds and so were more assured about the long-term prospects of nongovernmental enterprises.

Technology entrepreneurs not only benefited from extensive social and political ties, but also from the lack of other organized interests able to influence local policy. The SOEs that remained in Beijing tended to be as conservative as their counterparts in Shanghai. In the late 1980s, for example, Stone approached a local SOE to be a partner in computer assembly. Stone offered the factory guaranteed employment and an increment of profit over five years in return for controlling power of the choice of products, management of the factory, and use of foreign exchange quota. These trade-offs were not negotiable in Beijing at the time, and Stone eventually located the factory in Yunnan.[16] It is likely that if the concentration of SOEs in Beijing had stayed high, managers in the state sector could have competed more effectively for many of the same resources needed by nongovernmental enterprises.

The Beijing Model of Innovation

Beijing's technological development strategy emerged from a number of sources. In October 1980 Chen Chunxian, a nuclear physicist from the Chinese Academy of Sciences, delivered a talk about his recent visit to the United States. Instead of speaking about basic research, however, he described what he had seen in Silicon Valley and around Boston. Chen later founded Advanced Technology Development Services, an organization many considered the first nongovernmental enterprise. Other academics and official study groups visited science parks in Silicon Valley, Boston, England, and Japan in the early 1980s.[17] Some individual entrepreneurs either spent time working in or studied the experiences of enterprises in California and Boston.[18] Moreover, policymakers and academics sponsored meetings with science and technology personnel to discuss how innovation could best be organized. As early as April 1984, the Haidian district gov-

---

16. Jici Wang and Jixian Wang, "An Analysis of New Technology Agglomeration in Beijing," *Environment and Planning: A* 30 (1998): 695.

17. See Chen Yisheng, "Zhongguo de Kexueyuan Yanjiu: 1984–1994" [Chinese science parks: 1984–1994], *Zhongguo Kuexueyuan Yuankan* [Report of the Chinese academy of sciences] 3 (1995): 237–40.

18. Zhao Mulan and Wang Delu, "Zhongguancun Erci Chuangye de 'Liu Hua' Zhanlue" [Zhongguancun's "six transformations" strategy], *Weilai yu Fazhan* [Future and development] 3 (1996): 14–16.

ernment held a meeting of more than twelve hundred S&T personnel to discuss the economic future of the district. The conference report declared that Haidian should become "China's Silicon Valley" and rely on high-tech enterprises to promote industrialization.[19] Later that same month, the City Science Commission held a meeting with eleven nongovernmental entrepreneurs to discuss the goals and objectives of technological development in Beijing.

Though no one overarching statement of the principles guiding Beijing's development exists, its components can be drawn from numerous research reports, planning documents, and local newspaper reports. Specifically, these reports address three questions: (1) what type of individuals should be the main force in technological development; (2) how should they organize for innovation; and (3) what was the role of the government in supporting technology industries? In Beijing, consensus answers to these questions have gradually emerged: individuals with high levels of educational achievement and technological know-how; these entrepreneurs should be encouraged to set up their own nongovernmental enterprises with growing links to other scientific units; and the local government should act to maintain and support autonomous enterprises.

The best account of the Beijing model comes from a series of articles published in the *Beijing Daily* about Zhongguancun, the center of Beijing's technological growth. During the 1980s, *minying* enterprise growth was increasingly located on Zhongguancun, "Electronics Avenue" (*dianzi yi tiao jie*), in Beijing's Haidian district. In 1988 the central government convened a working group made up of seven departments of the local and central government to produce an investigative report on the street. The report summarized the basic experiences of the street, analyzing the reasons for its success and suggesting a framework for continued local government support.[20] In March and April of the same year, *Beijing Daily* published a six-part series on the report and "Electronics Avenue." These articles, more than any previous government regulation, formalized the meaning of nongovernmental enterprises in the local economy. Central to the district's success, according to the articles, was the technological and commercial abilities of the individual entrepreneurs themselves. The central and local government promoted the policies that allowed S&T personnel to found new enterprises, but it was these individuals, working through the institution of the technology enterprise, that had been behind the success of the district.

While the experience of nongovernmental enterprises was contrasted favorably against SOEs, ownership type was not central to their definition.

19. "Kaichuang Zhongguoshi de Guigu de Tansuo" [Exploration of building a Chinese-style Silicon Valley], *Beijing Ribao,* 11 September 1984, 1.

20. "'Zhongguancun Dianzi Yi Tao Jie' Diaocha Baogao" [Report on the investigation of Zhonguancun's "electronic avenue"], *Renmin Ribao* [People's daily], 12 March 1988.

Rather management strategies and business styles were.[21] Nongovernmental enterprises were self-financed and self-managed. New enterprises divided strategic and day-to-day management responsibilities between a board of directors and a general manager.[22] Start-up capital was not drawn from the plan or from the state financial distribution system; enterprises raised money from a variety of sources including communes (Stone), the Chinese Academy of Science (Legend), and universities (Founders). Once established, these enterprises operated within a new, more competitive economic sector that was "market led." Market pressures broke down the barriers within institutions; within enterprises, research, marketing, and sales departments cooperated in the development of new products.

Moreover, the articles stressed that unlike in the planned economy, a dense network of ties connected enterprises in the high-tech sector across different levels of different administrative bureaucracies. Collective, state, research, and university organizations had extensive ties to each other and worked together. Later the local government trumpeted interorganizational cooperation and entrepreneurial activity as the basis of an "innovation network" (*chuangxin wangluo*) linking enterprises, research institutes, and universities to each other and to the local government.[23]

While these ties were important, the report also stressed that each enterprise had to make it on its own. Not knowing how technological markets were going to develop and which products were going to be popular meant failure. Unlike SOEs, nongovernmental enterprises may actually go bankrupt. Yet entrepreneurs and administrators both saw the ability to fail as a sign of success for the city.[24] As an official at the China Nongovernmental S&T Entrepreneurs Association put it, "We [the government] are going to let the market provide signals, and if *minying* enterprises fail, well, then they fail."[25]

Underlying this model of development was, for the Chinese political context, a relatively limited view of the role of government. Enterprises were the products of a "large and a small reform environment." The central government, through the reforms of the state science and technology system, created the large environment. The actions and policies that the city and district governments enacted to support new enterprises produced a small environment. In particular, newspaper articles mentioned the sup-

21. Liu Hong, Ning Ming, and Wang Jianhua, "Zhongguancun Dianzi Yi Tiaojie de Qishi (si)" [Zhongguancun's electronics avenue: Part 4], *Beijing Ribao*, 19 April 1988.

22. Liu Hong, Ning Ming, and Wang Jianhua, "Zhongguancun Dianzi Yi Tiaojie de Qishi (yi)" [Zhongguancun's electronics avenue: Part 1], *Beijing Ribao*, 28 March 1988.

23. Zhongguancun Innovation Potential Discussion Group, "Peiyu Zhongguancun Quyu Chuangxin Wangluo, Zengqiang Chuangxin Nengli, Cujin Gaoxinjishu Chanye Fazhan" [Cultivate Zhongguancun district's innovation network, strengthen its innovative ability, and accelerate high-tech industrial development], in *Beijingshi Xinjishu Chanye Kaifa Shiyan Qu Yanjiu Baogao: 1995* [Research report on the Beijing Experimental Zone: 1995] (Beijing, 1996), 7–25.

24. One official noted, not without a little pride, that between 1989 and 1993 200 enterprises failed a year. Three hundred new ones quickly replaced them. Interview, B15, 24 April 1997.

25. Interview, B31, 7 July 1997.

port of the tax, labor, and commerce departments of Haidian district.[26] Reports in the *Beijing Daily* argued that recognizing and registering high-tech enterprises was essential, for "without the support of the district government, there would be no Electronics Avenue."[27] The reports also argued that Zhongguancun succeeded because both levels of governments shifted their roles from direct planning to macro-coordination; they had reduced government intervention and placed enterprises and markets at the center of the development plan.[28]

Exemplifying this shift was the establishment of a coordinating committee (*xietiao weiyuanhui*) made up of the Beijing city government, Haidian district government, and representatives of the supervisory organizations that had established enterprises in the district. In sharp contrast to the role other local governments played in the development of specific industrial sectors, the committee's role was extremely limited.[29] The committee's mission was to set long and short-term goals for the district, coordinate the flow of information between enterprises and the local government, and to provide services like market research or export promotion to enterprises. The committee was also to encourage S&T personnel to set up their own companies, explaining the opportunities available to scientists and helping them navigate the bureaucratic hurdles to registering a new enterprise. In Chinese, the committee's role is described as *guihua,* or creating a broad framework for development, which identifies general orientation and defines the focus of the industry over a period of several years, as opposed to *jihua,* the concrete planning of specific goals.

Government action was to distinguish between inside and outside the enterprise. Inside the enterprise, the local government should have no role. Outside the enterprise, officials should act to protect the internal working and autonomy of enterprises. According to Vice Mayor Hu Zhaoguang, success "depends on whether the local government can protect the internal operations of the enterprise."[30] This often meant protecting enterprises from interfering and extractive agents. In some cases, given the unevenness of the reform process, the local government would have to take a more activist role. But in these instances, the enterprise was to re-

26. "'Dianzi yi tiao jie' Wei Keji Gaige Tigong Chenggong Jingyan" ["Electronics avenue" is a success in reforming science and technology], *Beijing Ribao,* 11 March 1988.

27. "'Zhongguancun Dianzi Yi Tao Jie' Diaocha Baogao" [Report on the investigation of Zhonguancun's "electronics avenue"], *Renmin Ribao* [People's daily], 12 March 1988.

28. Report discussed in "Beijing Xinjishu Chanye Kaifaqu Yao Jiduan Pandeng Xingaoshan [Beijing experimental zone must continue climbing new heights], *Zhongguo Keji Luntan* [China S&T forum] 3 (1994): 43–47.

29. See, for example, the discussion of the Shanghai Automotive Industrial Corporation in Eric Thun, "Changing Lanes in China: Reform and Development in a Transitional Economy" (Ph.D. diss., Harvard University, 1999).

30. Before becoming vice mayor of Beijing, Hu was director of the BEZ and district mayor of Haidian. Quote is from Hu Zhaoguang, "Dui Beijing Xinjishu Kaifa Shiyanqu de Lixing Sikao" [Some thoughts on the Beijing experimental zone], *Guowai Keji Zhengci yu Guanli* [Foreign S&T policy and management] 1 (1993): 1–4.

main at the center of development, and local government activity was to reproduce market pressures.

The distinction between inside and outside the enterprise, however, occasionally blurred. Government activity sometimes fell short of the ideas delineated in the *Beijing Daily* article. But these ideas motivated and structured city government actions in three areas: financial, property rights, and government supervision. Within these three areas, specific policies like defining property rights directly influenced internal management styles. Other policies more indirectly influenced governance structures. In either case, the local government sought to strengthen its conception of nongovernmental development and duplicate it in as many technology enterprises as possible.

THE ORGANIZATION OF NONGOVERNMENTAL ENTERPRISES

The focus on government policy should not suggest that entrepreneurs simply reacted to local government initiatives. The strength of will and capacity for hard work of the men and women who began these enterprises cannot be underestimated.[31] Often described as fearless (*danzi hen da*), these were people who had risked the safety and benefits of high-level jobs in the state S&T system. They had a vision of their role in society and they expected to achieve their goals within the new enterprises. In one interview, the founder of one of the first *minying* enterprises called Jinghai, Jiang Shifei, described why he had left CAS. "If it is money I wanted," he noted, "I would have stayed abroad." Instead, establishing Jinghai allowed him to enter "an environment in which I could develop myself . . . where it doesn't matter what experience you have, as long as you have ability."[32] Clichéd as it might sound, entrepreneurs believed they were affecting change at both the personal and national level.[33]

Rather than the local government dictating the terms of development, the process of enterprise creation was much more one of mutual feedback. By the late 1980s entrepreneurs began discussing and writing about many of the barriers faced by growing enterprises and how these problems should be addressed. In the early 1990s, for example, managers at Stone began discussing how they had reached a new stage of development, "the

31. In contrast to those employed in SOEs, workers in *minying* enterprises were seen to be extremely industrious. One book reported that 40 percent of *minying* entrepreneurs worked longer than eight hours a day, 48 percent ten hours or longer, and 17 percent over twelve hours a day. Yan Zulin, *Xin Jishu Kaifaqu yu Keji Qiye Fuhuaqi* [New technology development parks and technology enterprise incubators] (Beijing: Kexue Chubanshe, 1991), 56.

32. "Renjin Qicai de Da Wutai" [Entering a unique stage], *Beijing Ribao*, 19 April 1988.

33. Prospectuses for nongovernmental enterprises often declare that they are "dedicating themselves to the service of China" (Stone) and "revitalizing the Chinese nation" (Legend). At the level of the individual, the resonance with Silicon Valley entrepreneurs is striking. "With all the talk of making money in the [Silicon] Valley, there is just as much verbiage about 'making a difference.'" "Moxie and Money: Silicon Valley, How It Really Works," *Business Week*, 18 August 1992.

second breakthrough" (*er ci chuang ye*). During this stage, nongovernmental enterprises would move from the "four self principles" (*si zi*) to the "six changes" (*liu hua*). As enterprises developed, they needed to globalize (*guoji hua*), move to stock systems (*gufen hua*), increase scale of production (*guimo hua*), become conglomerates (*jituan hua*), industrialize (*chanye hua*), and financially diversify (*jinrong dou yuan hua*).[34] Other enterprises, newspaper reporters, and think tanks picked up these ideas, and eventually they were reproduced in local government reports and policy pronouncements. Vice Mayor Hu Zhaoguang noted in 1995 that, like Stone, the Beijing Experimental Zone (BEZ) had entered the second stage of development and the local government's role needed to change.[35]

A process of interaction, however, should not be assumed to be one without discord. Local government and economic actors did not always have the same views about how to organize new enterprises, and there were conflicts about what the local government should and should not do for nongovernmental enterprise. The barriers to more congenial cooperation were likely to be political as well as economic. As Wan Runnan of Stone put it, "To be a successful entrepreneur means not only conflicting with the old S&T system, but also to battle with old concepts."[36]

As a whole, *minying* enterprises have gone through two stages of development: 1980–1992, and 1993–2000. The first stage was one of rapid growth and then consolidation. By 1993, the conditions of the domestic market as well as the government's approach to technology enterprises had changed dramatically. High-tech entrepreneurs could no longer expect to maintain the rapid growth rates of the 1980s and had to adapt to market changes. Future success would rely on significant changes in enterprise organization—the second breakthrough.[37] The tax breaks and tax holidays that sustained many of the first nongovernmental enterprises were eliminated in 1993. Also by this time, CAS ended support for many of the more established enterprises, shifting its focus and support to new ventures.[38]

---

34. See "Sitong Jituan" [Stone Corporation], in Beijing City Science and Technology Commission, *Pouxi Xianzhuang, Yingzao Weilai: Zhongdian Gaoxin Jishu Qiye* [Analyzing the present, building the future: Key new and high-technology enterprises] (Beijing: 1997).

35. Xie Ning, "Beijingshi Fushi Zhang Hu Zhaoguang Tan: Beijing Xinjishu Chanye Kaifaqu Fazhan Zhanlue" [A lecture by Beijing vice mayor Hu Zhaoguang: BEZ's development strategy], *Xinxi Shidai Daokan* [Contemporary information technology report] 3 (1993): 4–7.

36. "Mubiao: Zhongguo de IBM: Ji Sitong Jituan Zongjingli Wan Runnan" [Objective: China's IBM: An interview with Stone's general manager Wan Runnan], *Beijing Ribao*, 14 March 1988.

37. This section draws on Fang Xin, "Gaojishu Xiaoqiye yu Chuangxin Yanjiu" [High technology, small enterprises, and innovation] (Ph.D. diss., Qinghua University, 1997).

38. The shift to new enterprises may have also been caused by the increasing stridency with which enterprise managers were protecting their territory (and profits) from CAS. In one interview, a manager admitted that all the technology the enterprise had commercialized had come from CAS, but that should not give CAS any claim to payment: "We are the ones that changed technology into a product." Interview, B34, 16 July 1997.

In addition, enterprises in Beijing faced an ever-growing array of competitors. Some SOEs became more flexible and entered the computer market. Competitive *minying* enterprises emerged in new regions like Hangzhou and Shenzhen. Foreign enterprises like IBM, Microsoft, and Fujitsu arrived in China, set up research centers, and entered new market areas like Chinese-language software that had been the exclusive domain of domestic producers. Many of these foreign firms attracted the brightest and most-talented scientists away from domestic enterprises with high salaries and opportunities to go abroad.

Growing competition was felt among domestic enterprises. Income and growth were increasingly distributed among a larger number of enterprises in Beijing. The most dramatic case has been the rise and (relative) fall of Stone. During the 1980s, Stone dominated the domestic market in Chinese-based word processors, holding some 80 percent of the market.[39] A failure to appreciate that Chinese-language software and personal computers would make word processors obsolete meant that Stone's income did not keep up with the domestic market. From 1993 to 1997, Stone's annual income little more than doubled (RMB 3.2 billion to RMB 6.4 billion); in the same time, Legend's income almost quadrupled from RMB 3.2 billion to RMB 12 billion. More generally, in 1990, the three top enterprises, Stone, Jinghai, and Xintong, accounted for 80 percent of all income earned by *minying* enterprises in Beijing.[40] By 1995, Founders and Legend replaced Jinghai and Xintong, and the percentage of total income earned by the top three fell to 54 percent.

FINANCIAL ACTIVITIES

The lack of venture capital was and continues to be the biggest barrier to growth in the high-tech sector. Most enterprises opened shop with limited capital raised by their supervisory agency, and a scarcity of other funds prevented even the most successful enterprises from increasing their scale of production and exploiting economies of scale. At times, the methods adopted by the local government to raise capital worked at cross-purposes to the goal of supporting independent and competitive enterprises. Some types of funding rewarded enterprises with political connections, not the most innovative products or processes. But overall, the local government tried to re-create, or at least not stifle, market pressures through its funding policies.

39. Scott Kennedy, "The Stone Group: State Client or Market Pathbreaker," *China Quarterly* 152 (December 1997): 746–77.

40. Beijingshi Kexue Jishu Weiyuan Hui, *1997 Niandu Beijing Keji Qiye Gongzuo Yaoloan* [1997 overview of Beijing technology enterprises] (Beijing: Zhonguo Jingji Chubanshe, 1998), 69.

In the face of severe fiscal constraints, the local government played both a direct and indirect role in the financing of technology development. Indirect funding came in the form of infrastructure projects or large-scale technology projects. More direct funding came from state-owned banks, the local government budget, central government disbursements under local government control, or more informal sources of capital. In addition, the explosion of internet technologies and dot-com companies in 1998 and 1999 attracted large inflows of foreign capital, but after the fall of the Nasdaq and the economic slowdown in the United States, the majority of nongovernmental enterprises returned to their original state of being investment capital poor.

Bank Lending

Nongovernmental enterprises were not officially recognized at the national level until 1985, and so were ineligible for funds from the central plan as well as bank loans. Unable to turn to banks, branches of the municipal government, often in coordination with the enterprise's supervisory unit, arranged funding from third sources. Haidian Science Commission officials, for example, arranged for communes or SOEs in the district to invest in *minying* enterprises.[41] In most cases, enterprises either relied on the capital the founding individuals could raise among themselves, or the initial investment of the supervisory agency. The one time loan of RMB 200,000 that CAS made to Legend plays a key part in both enterprise literature and newspaper accounts of the enterprise.

After the 1985 Decision Concerning the Reform of the Science and Technology Management System, *minying* enterprises could officially apply for bank loans. But the decision did little to change the fact that local governments directed bank lending and ordered banks to loan to SOEs. Faced with unemployment and the other social ills that bankrupt SOEs would create, local governments, even in Beijing, rarely diverted bank lending to nongovernmental enterprises. In this context, local officials sometimes did their own fundraising, acting as matchmakers between entrepreneurs and more informal capital sources. In Chaoyang district, for example, the head of the science office sought out his own friends to raise a RMB 200,000 investment when an enterprise in the district could not get needed capital.[42]

Even when banks were allowed to make independent loans based on po-

41. The most famous would be the commune that invested in Stone. Interview, B2, 5 May 1996.

42. Zhang Yi, "Minying qiye de 'hao popo'" [A "good mother-in-law" to nongovernmental science enterprises], *Zhongguo Minying Kexue yu Jishu* [Chinese nongovernmental science and technology] 6 (1997): 39.

tential returns, a nongovernmental enterprise was not the borrower of first choice for the state-managed banking system. It was unclear which, if any, government actor could guarantee the loan. Moreover, though some *minying* enterprises would eventually have high rates of profit, many others went bankrupt. In some cases, branches of the local government acted as an intermediary in order to secure a loan for *minying* enterprises. For example, Haidian district government officials helped arrange a loan for Chen Chunxian, the first *minying* entrepreneur and a research fellow in the Institute of Physics at the Chinese Academy of Sciences, in 1986 from the Haidian Agricultural Bank.[43] Success in securing bank funds, however, depended on either political connections or a reputation as a successful enterprise, and access to capital remained extremely tight. A 1993 survey of 213 enterprises within the Beijing Experimental Zone revealed that only 17, or 7 percent, received funding from a bank.[44] Only the largest and most established enterprises used the banks; start-up or unconnected enterprises were still excluded from bank lending.[45]

During the "second breakthrough," securing capital became both an easier and more daunting process. Easier because in 1993 the SSTC and the Reform Commission issued the Decision Regarding Several Problems in Promoting Nongovernmental Technology Enterprises.[46] Officially recognized at the national level, *minying* enterprises could now more easily approach banks for loans. Moreover, the past decade of rapid growth in nongovernmental enterprises attracted the attention of other nontraditional investors, including SOEs. Faced with reduced budgets and large staffs, some SOEs looked to high-tech enterprises as a source of easy revenue. As the founder of an internet company who used a loan from an SOE put it, "We do not use them (the SOE) for political protection. They use us because they need the money."[47]

At the same time, it became harder for most nongovernmental enterprises to secure loans. The reform of the banking system converted banks into commercial (*shangye*) ventures; they were no longer to act as the disbursement arm of the central plan but rather focus on returns on loans.

43. Though it should be noted this was five years after he had founded his company. Zheng Haiding, "Kaiming de Popo" [Enlightened mother-in-law], *Keji Ribao* [Science and technology daily], 19 February 1987.

44. Keji yu Jinrong Gaige Ketizu [Study group in science and technology and finance reform], "Jinrong Tizhi Gaige Dui Gaoxinjishu Qiye de Fazhan Yingxiang ji Duice" [The impact of financial reform on high-tech enterprise development and some countermeasures], in *Beijingshi Xinjishu Chanye Kaifa Shiyan Qu Yanjiu Baogao: 1995* [Research report on the Beijing Experimental Zone: 1995] (Beijing, 1996), 16.

45. Of the enterprises interviewed in Beijing, no start-up funds had come from a bank.

46. "Guanyu Dali Tuidong Minying Keji Qiye Fazhan Ruogan Wenti de Jueding" [Decision concerning the problem of vigorously promoting nongovernmental technology enterprises] in *Keji Fagui Xuanbian* [Selected S&T laws and regulations] (Xi'an: Xi'an Kexue Jishu Weiyuanhui, 1996), 390–98.

47. Interviews, B20a, B20b, 14 May 1997.

Declining rates of return in the computer industry made nongovernmental enterprises less attractive borrowers. Banks focused either on parts of the economy with higher rates of return, like real estate, or less risk, like SOEs that did not face the real possibility of going bankrupt.

Science Budgets

Infrastructure projects and investment in the local science budget were the easiest ways to ensure the reproduction of the nongovernmental ideal. These investments were not enterprise specific and their benefits were not available to enterprises only on the basis of political or administrative connections. In 1991, the State Science and Technology Commission and the Beijing municipal government began construction of the Shangdi Information Industry Base.[48] Shangdi, located six kilometers north of Zhongguancun, was China's first zone dedicated to developing information technologies.[49] In addition, the local government doubled its technology budget during the Eighth Five-Year Plan (1991–1995) to $1.3 billion. The city borrowed $700–$800 million from foreign governments and the World Bank, $200–$300 million from foreign commercial banks, and the rest from domestic banks.[50] In 1994, Beijing invested over RMB 10 billion in scientific research and development organizations.[51] Overall investments in S&T activities were twice as large during the Ninth Five-Year Plan (1996–2000) as they were during the Eighth.[52]

One of the advantages for Beijing was how S&T investments were actually disbursed. Unlike in Shanghai, the majority of the city's research funds went to independent scientific organizations, not to labs within SOEs.[53] In 1992 for example, only 14 percent of the city's S&T appropriations went to medium and large-sized SOEs, while 58 percent went to research institutes.[54] Such funding not only ensured that research funds were not diverted to welfare concerns within the enterprise, but also promoted horizontal links between production and research units. By promoting these links, and by creating a market in which enterprises with the best products can exploit these links, the local government used broad-

48. Interviews, B35, 17 July 1997, and B24, 26 May 1997.

49. Office of the Beijing Experimental Zone for the Development of New Technologies, "Beijing City Experimental Zone for the Development of New Technologies: Investment Guide," n.d.

50. *China Daily*, 17 August 1992, 2.

51. *Zhongguo Keji Tongji Nianjian: 1995* [China statistical yearbook on science and technology: 1995] (Beijing: Zhongguo Tongji Chubanshe, 1995), 4.

52. "2000 nian Beijingshi Guomin Jingji he Shehui Fazahan Tongji Gongbao" [2000 Beijing economic and social development statistics], at http://www.stats.gov.cn/tjgb/beijing/200103090129.htm, accessed 8 September 2001.

53. In 1994, Shanghai invested six times as much in research activity in large- and medium-sized enterprises than Beijing did. *Zhongguo Keji Tongji Nianjian: 1995*, 39.

54. *Beijing Keji Tongji Nianjian: 1992* [Beijing statistical yearbook in science and technology: 1992] (Beijingshi Kexue Jishue Chubanshe, 1993), 107.

based funding to ensure that competition shaped interactions between *minying* enterprises.[55]

Direct Investment in Specific Enterprises

Arranging direct funding for enterprises was one of the most efficacious ways for the local government to support individual enterprise development, yet it ran counter to the goal of creating an open and competitive market. Direct funding by the municipality may have gone to enterprises with the best political connections or to enterprises with a reputation for success, not to the enterprises with the most competitive products for the current generation of development. Evidence for direct investment by the Beijing municipality is mainly anecdotal at the beginning stages of reform. Until 1994, the local government could use a portion of the tax revenues collected in the zone to make loans to enterprises. In 1993 these funds totaled RMB 300 million and were referred to as a "turnover" or "revolving" (*zhouzhuan jin*) fund. The BEZ, for example, provided a loan of RMB 200,000 to YaDu, an enterprise that had registered with only RMB 500,000 in total assets.[56]

In 1994, central tax reforms eliminated this fund, but the local government still possessed some ability to invest directly (or issue its own loans) to enterprises. Two 1994 graduates of Qinghua University wanted to develop a new touch-screen technology. Unable to raise the funds themselves or convince their research institute to invest in the project, they approached the science commission of Haidian district, which agreed to invest RMB 45,000.[57] How prevalent this was in Beijing is unknown; nationally, local governments accounted for only 5 percent of investment in high-tech enterprises in 1990, and 9 percent of enterprises in the Beijing science park list the city as their supervisory unit.[58]

In order to be able to direct funding to promising enterprises, the city

55. The question of how enterprises link up with research labs, or, for that matter, how they make sales is extremely complicated in China. How important a role does personal connections, or *guanxi*, play?

When the question is put bluntly (and rather badly), entrepreneurs insist that *guanxi* has played no role in their success. When the question is followed up by some statement describing the way the enterprise has used a connection in the past, the respondent often says *guanxi* is important, but not in the same ways it used to be. For example, when I pointed out that a software company's largest customer was the founder's old work unit, the interviewee conceded that connections had been important. But he also stated that connections had only "opened the door. If the product had been no good, *guanxi* would be of no use." Connections, however, appear extremely important in the arranging of bank loans.

56. Zaishui Tizhi Gaige Ketizu [Research group on tax reform], "Zaishui Tizhi Gaige yu Beijing Shiyanqu Gaoxinjishu Qiye Fazhan" [Tax reform and high-technology enterprise development], in *Beijingshi Xinjishu: 1995* [Research report: 1995], 6.

Interestingly, Microsoft later sued YaDu for using pirated versions of its software, and the case generated much anti-Microsoft sentiment in China. See "Microsoft Sues Chinese Enterprise for Using Pirated Software," Agence France Presse, 19 November 1999.

57. Interview, B14, 7 April 1997.

58. Shilin Gu, *Spin-off Enterprises in China: Channeling the Components of R&D Institutions into Innovative Business*, UNU/INTECH Working Paper No. 16, December 1994, 14.

government set up its own technology fund in May 1999. The Beijing Technology Development Fund, financed with $10 million from the Beijing municipal government and $40 million from W. I. Harper Group, a U. S. institutional investor, will focus on software, internet, telecommunications, and biotechnology. By 2001, Beijing had established four other funds, bringing total funds available to $120 million, far short of the development needs of enterprises. By comparison, Singapore established in 1999 a $1 billion fund to sponsor what it calls "technopreneurship." Moreover, government funds are still driven by policy, and government agencies are often unable and unwilling to accept the high risk of technology investment.

The Torch Plan

Initiated by the central government in May 1988, the Torch Plan sought to broaden the sources of funds available to nongovernmental enterprises. Local science commissions were to determine how Torch Plan funds were spent, and in Beijing the local government used Torch funds to support new start-ups and some of the larger enterprises, including Stone, Legend, Founders, and Jinghai.[59] The effect of these funds is hard to measure and seems to be determined by the enterprise's stage of development. Since most Torch grants were small, larger enterprises needed to secure additional capital from other sources. For both large and small enterprises, being included in a Torch Plan project acted like a seal of approval making it easier for them to gain funding from other sources.

As part of the Torch Plan, banks within the Beijing Export Zone were directed to loan to *minying* enterprises. But even with the plan and the legal recognition of *minying* as a property type, enterprises had difficulty securing loans without the intervention of BEZ officials. Under these conditions, the BEZ management office used political pressure to convince local banks to make loans.[60] The BEZ did not flex this muscle without extracting something in return from the enterprises. In order to be eligible for the loans, enterprises were expected to display a higher degree of transparency than typically found in Chinese enterprises. Moreover, internal management and accounting were to attain the "scientific and modern level" set by the management office. If they could not meet these levels, experts from the BEZ would work with the enterprise to assure they did.[61]

Financial Constraints and Enterprise Behavior

Tight capital constraints produced a fairly consistent pattern of behavior among nongovernmental entrepreneurs across China. At the very begin-

59. Interview, B16, 27 April 1997. In the brochure describing the Torch Plan, on the page after a picture of Deng Xiaoping's inscription on high technology, are photos of Stone, Founders, and Legend corporations. Guojia Kewei Huoju Gaojishu Chanye Kaifa Zhongxin [SSTC Torch high-technology industry development center], *Huaju Jihua* [China torch program] (Beijing, n.d.).

60. Interview, B17, 5 May 1997.

61. Interview, B15, 24 April 1997.

ning, most spin-off enterprises received their capital from the units they spun off from; Legend began with a loan of RMB 200,000 from CAS; Stone with private savings and a loan from the commune that was its official sponsor. Of 271 enterprises surveyed in the BEZ in 1995, 76 percent of those enterprises identified as state owned either borrowed or were allocated their start-up capital from an official agency. Seventy-five percent of collective enterprises that registered with a supervisory agency also borrowed or were allocated funds from an official agency. Enterprises started by individuals, or groups of friends, tended to rely on personal savings or loans from family members; 63 percent of collective enterprises without a supervisory agency listed personal savings as the source of their start-up capital.[62]

With other routes to finance cut off, nongovernmental enterprises relied on income generated from sales. During the 1980s, *minying* enterprises benefited from being the first to enter the market with new products. Stone dominated the Chinese word processing market with the MS-2400 and subsequent generations of the same product, controlling over 80 percent of sales. Founders and Legend also had significant technological leads on their competitors in Chinese-language publishing software and character recognition processes respectively and could rely on large incomes during the 1980s. Moreover, they also benefited from rising living standards in urban areas and increased urban demand sparked a "computer fever" in the mid 1980s. As a result total *minying* enterprise income within the city multiplied ten times, from RMB 1.4 billion to RMB 14 billion, between 1988 to 1994.

Many businesses also focused on the sales of Western products. All the large enterprises had representative agreements with enterprises such as IBM, Compaq, Apple, and Hewlett-Packard. Early agreements with foreign companies were a key ingredient to later success. Begun in the 1980s, distribution agreements with MNCs provided *minying* enterprises with much-needed investment capital, and estimates of the value of such sales for *minying* business varied from 20 to 70 percent.[63] Trade remained a significant part of revenues until the early 1990s.[64] Only the largest and most innovative enterprises (a small percentage of the total number of *minying* enter-

62. Keji yu Jinrong Gaige Ketizu [Study group in science and technology and finance reform], "Jinrong Tizhi Gaige Dui Gaoxinjishu Qiye de Fazhan Yingxiang ji Duice" [The impact of financial reform on high-tech enterprise development and some countermeasures], in Beijingshi Xinjishu Chanye Kaifa Shiyan Qu Xiezuo Xiaozu (BEZ small writing group), *Beijingshi Xinjishu: 1995* [Research report: 1995], 28.

63. This probably differs by scale of product, even within the broad sector of information industries. The 20-percent figure comes from the manager of a mainframe computer company involved in railroads (interview, B30, 7 July 1997), 70 percent from a researcher involved in the PC market (interview, B12, 19 March 1997).

64. For enterprises located within the national high-tech development zones, the percentage of income from trade decreased from 28 percent in 1989 to 13 percent in 1997. Sales income was 81 percent. This data is not broken down by enterprise type or by region. See *China Torch Program Statistics for Ten Years,* Administration Office Torch Program, Ministry of Science and Technology, PRC, n.d., 14.

prises) have been able to gradually decrease that reliance; within the Beijing Experimental Zone, the share of income derived from manufacturing increased from 31.4 to 37 percent between 1991 and 1994.[65]

In 1992 the central government began to lift import quotas on foreign personal computers and substantially reduced tariffs. Increased competition and the arrival of the foreign firms themselves meant that nongovernmental enterprises could no longer rely on representative agreements to raise cash. Instead, many enterprises rapidly diversified into new markets in the search of quick gains. Founders moved from publishing software to computers, chemical engineering, and real estate. Stone expanded from word processors to real estate, cement, and chocolate snack pies. In the words of one Founders manager, these changes were the reflection of the shift from a "product" to a "market" company: "Before we produced what we thought we could sell; now we research what the market wants to buy."[66] While such a strategy may be successful in raising capital quickly, it also risks moving the company farther and farther away from its core technologies.[67]

Diversification has also occurred into the internet and web technologies. Legend shifted its unprofitable motherboard production to Shenzhen, shedding five hundred employees. Liu Chuanzhi, chairman of the Legend Group, has repeatedly stated that the internet is Legend's future, and in October 2000 Legend announced the founding of Digital China Corp. Ltd. to signal its entrance into the field of e-commerce. Digital China will provide e-commerce and systems planning. Both Founders and Stone have also entered the market, with Founders moving away from typesetting and publishing systems to more internet-based technologies.

Both companies have also pursued a strategy based on joint ventures.[68] Legend formed a joint venture in producing Toshiba notebooks, launched a software venture with Computer Associates, and in June 2001 announced an alliance with AOL. Stone set up major joint ventures with Fujitsu, Mitsui, Mitsubishi, and Compaq. Cooperation with these companies provided access to technology, global distribution networks, management "know-how," and capital.

The largest and richest enterprises have also set up their own investment arms. Legend has a research wing that investigates and makes small invest-

65. Liu Dong, "Beijing Minying Keji Qiye Xianzhuang he Qushi" [Research into the current situation and future trends of Beijing nongovernmental enterprise development] (unpublished ms., n.d.).

66. Interview, B39, 29 July 1997.

67. Again this varies by enterprise. Stone appears to be more at risk than other companies. This may account for most respondents' belief that the next five years were going to be rough, and that the company was desperately seeking a new product. According to a respondent at Founders, information industries still account for 60 to 80 percent of revenue.

68. Wenlei Li and Jasper Yang, "Chinese Home-grown Names: Case Studies of Legend and Stone," *Harvard China Review* 1, 1 (summer 1998). Online at http://www.harvardchina.org/magazine/article/home-grown1.html, accessed 15 June 2000.

ments in other companies. By 1996, they had made investments ranging from RMB 3 million to RMB 5 million in twelve companies. Legend has also started construction on an "electronics city" in Huiyang County, Guangdong. It hopes to collect rental and management fees from domestic and electronics companies that relocate to the site.[69] In April 2000 Legend announced its intentions of establishing a RMB 3.8 billion venture capital fund.[70]

Finally, the internet boom also meant the arrival of foreign venture capital funds. Many of these firms were based in Silicon Valley and hoped to link Chinese working in California to entrepreneurs in Zhongguancun. Beijing was reported to be the most active market for venture capital in China, with one survey claiming the city had attracted $700 million in venture capital.[71] But much of this money appears to be "bubble" funds based on unrealistic expectations of internet growth in the United States and China; actually realized venture capital only reached $145 million by the first half of 2000 when the internet bubble burst.[72]

More importantly, the long-term policy environment for venture capital funds remains uncertain. The People's Bank of China bans foreign venture funds, but allows them to establish representative offices. Restrictive policies also limit exit strategies; few start-ups are allowed to list on the domestic stock markets in Shanghai or Shenzhen, much less on markets abroad. Direct initial public offerings are not possible, and the law still does not allow for preferential or convertible shares of stock. The central government has begun steps to address the absence of appropriate legislation; draft regulations were expected to be submitted to the National People's Congress before the end of 2001.

PROPERTY RIGHTS REFORM

Property rights reform has been extremely complicated in Beijing and, like the rest of the reform process, characterized by starts and stops. In the early stages of reform, supervising and regulating new institutions provided stability and legitimacy to nongovernmental enterprises. At later stages, officials in local government branches encouraged changes in ownership structures, or quickly moved to legitimate experiments made by the high-tech entrepreneurs themselves. Local officials gradually focused on defining, refining, and institutionalizing the concept of nongovernmental enterprise. Unlike in Shanghai, they did not use property rights reform solely

69. Interview, B21, 21 May 1997.

70. http://chinaonline.com/industry/infotech/NewsArchive/Secure/2000/april/C000 41005.asp, accessed June 2001.

71. "Venture Capital Investors Favor Beijing," *People's Daily*, 19 February 2001.

72. "Realized Venture Investment Only Accounts for $145 million," http://www.china online.com/topstories/010620/1/c01061509.asp, accessed June 2001.

as a method to regenerate moribund state-owned factories. Rather, the main goal was to create a model of a modern enterprise.

The first *minying* enterprise was established in Beijing in 1980 when Chen Chunxian founded a technological consulting company. Chen's enterprise was quickly followed by a few other small ventures, most with very unclear boundaries between the ventures themselves and the sponsoring institution. Technology commercialized by the enterprise had frequently been developed in and was still owned by the supervisory agency. According to one report, 95 percent of all technology developed by *minying* enterprises in the BEZ originated from a state-owned unit.[73] Most employees continued to be simultaneously connected to their original units and the new enterprise. Within CAS, this dual employment was widely known as "one academy, two systems."[74]

During this early, transitional stage ambiguity in government relations and ownership structures had its benefits.[75] A more flexible structure of ownership allowed enterprises access to scarce resources, like technology and capital, from both the state and the market. Moreover, too strict a definition of property rights would have strangled innovation.[76] Since the state officially owned all technology, enforcing a clear definition of property rights would eliminate any incentives for individual scientists to commercialize new products. As one journalist put it, "Knowing who owns something is not always the best. CAS from 1950 to 1978 owned all the technology, and in all that time did not sell one product. Since the reforms, 40,000 products have passed to companies, and they have put the products on the market."[77]

Minying development stumbled along after the first enterprises were founded, especially in 1982 and 1983 after repeated ideological attacks on nongovernmental entrepreneurs. In July 1984 the city government issued its "Inquiry into Reforming the S&T System," which praised *minying* enterprises as an important engine of growth. The document also recognized collective (*jiti*) and individual (*geti*) forms of ownership as complements to the national public economy, legitimizing the property rights categories that *minying* enterprises tended to fall under at the time. The Commercial and Industrial Bureau began registering and regulating these enterprises.[78] The "Inquiry"

73. Chinese Academy of Science and Beijing Science and Technology Commission, "Beijing Xinjishu Kaifa Shiyanqu Xinjishu Qiye Yanjiu" [Study of new and high-technology industries in the BEZ], *Zhongguo Ruan Kexue* [Chinese soft science] 2 (1992): 40–44.

74. A play on Hong Kong's "one country, two systems."

75. Corinna-Barbara Francis, "Bargained Property Rights in China's Market Transition to a Market Economy: The Case of the High-Technology Sector," paper presented to American Political Science Association Annual Convention, Washington, D.C., 1997.

76. That is, ambiguity was better than state ownership, not the clear definition of property rights.

77. Interview, B36, 18 July 1997.

78. Beijing Minban Keji Shiyejia Xiehui, *Beijing Minban Keji Shiye Dashiji: 1980–1990* [A chronicle of Beijing's nongovernmental technology industry: 1980–1990] (Beijing: Zhonghua Gongshang Lianhe Chubanshe, 1994).

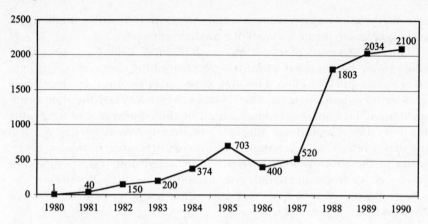

Figure 3.1. Number of Beijing *minying* enterprises, 1980–1990. *Source: 1997 Niandu Beijing Keji Qiye Gongzuo Yaloan* [1997 overview of Beijing technology enterprises], 67.

laid the institutional groundwork for and was a reaction to the explosion of enterprises that took place in 1984. By the end of the year, there were forty *minying* enterprises in Beijing, with an operating budget of RMB 18 million.

It is important that the local government in 1984 moved early to legitimize the legal position of *minying* enterprises. When local governments hesitated to recognize and define nongovernmental enterprises, as they did in Shanghai, Guangzhou, and Xi'an, entrepreneurs had difficulty securing access to funding and interacting with local bureaucracies. Recognition and regulation increased an enterprise's chance of survival. In the absence of effective tax supervision and legal protection of property rights, a partner in a high-tech enterprise often had the incentive and the ability to manipulate the internal books. Once discovered by the other partner, the enterprise often split into two smaller companies lacking the ability to compete. In fact, in a survey of 271 enterprises within the Beijing Experimental Zone, nongovernmental enterprises without a supervisory agency were the most likely to have a change in status of the prime investor.[79] Under these conditions, the management office of the BEZ acted as a guarantor, supervising the operation of the enterprise.

Nongovernmental Enterprises and Intellectual Property Rights
Moreover, local governments were the key actors in enforcing patent and intellectual property laws. During the Maoist years, technology was considered a public good; SOEs did not pay research organizations for the

79. Chanquan Zhidu Gaige Ketizu [Study group on property rights reform], "Beijingshi Yanqu Xinjishu Qiye Chanquan Gaige Diaocha yu Fenxi" [A report and analysis of property rights reform among BEZ enterprises], in *Beijingshi Xinjishu: 1995* [Research report: 1995], 27–40.

use of new technologies or processes. The reforms created a market value for transferred technologies, but some enterprises balked at paying for what they had once received for free. In one case where an enterprise refused to pay for transferred technology, Haidian district officials intervened, contacted the SOE, and when persuasion failed, pursued the case through the courts.[80]

Software piracy threatens innovative Chinese enterprises as much as it does foreign MNCs. In one survey of domestic software companies, 26 percent of the respondents cited piracy as the major obstacle to the development of the industry.[81] Most of the burden of enforcing the regulations rests on the enterprises themselves; they must collect all the evidence, including the name and addresses of all those involved in the criminal activity. But what district the enterprise is located in and the relationship between the enterprise and the local branch of the public security ministry often determines if local officials will pursue the case.[82]

Officials in local government branches did more than just recognize an emergent property structure; they also moved to legitimate experiments made by the high-tech entrepreneurs themselves. The Beijing City Reform Commission approved Stone's use of an internal stock system (*gufen hua*) in 1988, almost five years before similar systems were promoted at the national level.[83] In the same year, Founders and Legend established holding companies in Hong Kong.[84]

By 1992, research reports produced by the BEZ and policy analysts at CAS argued that stock systems would clarify ownership structures, modernize management, and increase investment in *minying* enterprises. The porousness of the boundaries between enterprise and supervisory unit that earlier supported enterprise growth had become a barrier since it was unclear who actually owned the company, office equipment, and even the commercialized technology. Enterprises without supervisory units frequently registered as collective when they entered the zone, even though an individual or a small group of friends had founded them. Known as "wearing a red hat" (*dai hong maozi*), registering as collective made sense politically given the still suspect nature of private property, but failed to recognize the role of the initial investor and venture capital.[85] The BEZ's

80. "Kaiming de Popo" [Enlightened mother-in-law], *Keji Ribao*, 19 February 1987.

81. "Study: Piracy Main Threat to China's Growing Software Industry," *China Online*, 8 September 2000.

82. These ties may be corrupt. After a Beijing entrepreneur described how he had spent three days collecting evidence of piracy, I asked him if the local judge pursued the case. His answer: "Of course. He had just come back from a holiday trip that I had paid for." Interview, B14, 7 April 1997.

83. "Xinjishu Chanye Kaifa Shiyanqu Fazhan Lianghao" [Experimental zone for the development of new technology industries develops well], *Beijing Ribao*, 20 November 1988.

84. "Sitong Gupiao Niandi Shangshi" [Stone to go on the stock market at the end of the year], *Keji Ribao*, 22 October 1988.

85. Ning, "Beijingsji Fushi Zhang Hu Zhaoguang Tan."

goal was gradually to convert enterprises that had originally registered in the zone as "collective" to "stock" companies. In 1992 less than 1 percent of all nongovernmental enterprises were using some type of stock system; by 1994 the number had risen to 10 percent.[86] Stock systems were viewed as the best way to separate ownership and management authority, as well as an effective way to create an incentive structure that would reward scientists for technological breakthroughs.

Property Rights Uncertainty and Enterprise Behavior

Organizationally, enterprises tried to cope with the ambiguity in property rights by creating a general manager and a board of directors. During the first years of operation in the 1980s, management was not much more than a problem of coordinating a group of close friends with a common background and similar goals. Everyday decisions, like hiring and firing, were officially the general manager's responsibility. But enterprises reflected the talents and energies of their founders and they often dominated the enterprises through the force of their personalities. The founders and their small group of friends frequently controlled decisions about business strategies, technology development, hiring, promotion, and research cooperation.[87] Staffed both by members of the supervisory unit and the enterprise, the board restricted itself to large investment and general strategy decisions. In the case of Legend, members of CAS, the Beijing local government, and the founders of Legend staffed the board. While CAS and the local government might have a say in the general direction of the enterprise, managers at Legend believed that the arrangement preserved their flexibility, for the "government is not able to interfere with internal management activities."[88]

As enterprises developed in Beijing, the relationship between employees and between the board and management grew increasingly complex. By 1993, the most successful enterprises were no longer small, although average enterprise size, 21 employees, remained smaller than in Shanghai.[89] Legend grew from 20 to 4,200 employees, Stone from 94 to 3,700. Com-

---

86. *1997 Niandu Beijing Keji Qiye Gongzuo Yaoloan* [1997 overview of Beijing technology enterprises] (Beijing: Zhonguo Jingji Chubanshe, 1998).

87. Even when the original entrepreneurs have said they have retired, some continue to run things behind the scenes. One former employee of Founders insisted that Wang Xuan continued to manage the company, down to decisions about who should get what apartments, even though he had retired from any official position within the company. He was still a professor at Beijing University and a member of CAS, as well as a delegate to the National People's Congress. Interview, B18, 8 May 1997.

88. Chen Huihu, *Lianxiang Weishenma* [Why Legend] (Beijing: Beijing Daxue Chunbanshe, 1997), 213.

89. For Shanghai, the average in 1994 was twenty-six. Shanghai numbers include part- and full-time employees. The same breakdown is not provided for Beijing. *Shanghaishi Keji Tongji Nianjian, 1995* [Shanghai S&T yearbook, 1995] (Shanghai: Shanghai Tongjiju Chubanshe, 1995),180.

panies could no longer rely solely on their supervisory agency for their staffing needs. Truly competitive enterprises had to attract the best employees from a range of sources, and employees were increasingly better educated and from various organizations. In 1992, S&T personnel made up about 24 percent of total employees in nongovernmental enterprises; by 1996 that percentage had risen to 61 percent.[90] When it started in 1988, for example, all Control Technique Development Group employees came from the Beijing Aeronautics Research Institute. By 1995, it drew a third from outside the institute through advertising and job fairs.[91]

New employees often had difficulty integrating into the existing organizational culture. Unlike the first generation of employees at Legend, newcomers were seen to be motivated by money, less willing to neglect their personal goals for the good of the enterprise, and more willing to pursue opportunities abroad or with foreign enterprises.[92] Moreover, new employees were mainly on salary and did not receive apartments or the other welfare benefits that older employees drawn from the original supervisory unit did.[93]

In order to balance the needs of these different generations, as well as to better react to increased market pressures, many enterprises began changing their internal management systems. Management based on personal authority no longer fit the needs of the growing scale of production.[94] Increasingly managers and directors at Stone, for example, came to believe "that one person having complete power leads to stagnation," and that the company needed to "rely on professional managers, not friends."[95] This increased professionalism of management was to reach across all areas of business operations. After 1993, operations were increasingly specialized and broken up into marketing, research, sales, and personnel divisions with special focus on R&D.[96] By 1995, 80 percent of all enterprises within the BEZ had independent research and development departments within

90. The 1992 number is from *Beijing Science and Technology Yearbook*, 1996 is from the 1997 report.

91. Interview, B30, 3 July 1997.

92. This reflected a growing age gap at the company, with few individuals bridging the two groups. Employees at Legend were either forty-six and above (40 percent), or below thirty (60 percent). Chen Huihu, *Lianxiang Weishenma* [Why Legend], 94.

93. Corinna-Barbara Francis, "Reproduction of the Danwei Institutional Features in the Context of China's Market Economy: The Case of Haidian District's High-Tech Sector," *China Quarterly* 147 (September 1996): 839–59.

94. These systems were frequently criticized as being "patriarchal" (*jiazhang zhi*).

95. Wan Runnan, "Guanyu Jiegou Sheji he Renshi Anpai de Shier tiao Yijian" [12 ideas about organizational design and personnel administration], internal speech, Stone Corporation, reprinted in *Zhongguo Minban Keji Shiyejia Xiehui Huixun* [Newsletter of the China nongovernmental science and technology entrepreneurs association], 21 January 1989, 2–3.

96. Companies that were *guoyou*, *minying* (owned by the state, run by the people), like Legend and Founders, also had a Communist party office. For other companies, the BEZ fulfilled that role.

the enterprise.[97] Legend, for example, had six departments, including offices for scientific development, finance, subsidiary companies, and production, as well as divisions for specific technologies including networks,
software, and microelectronics. Division heads were responsible for day-to-
day management decisions and reported to a general oversight office (*zong-
cai bangongshi*), which in turn reported to the board of directors (*dongshi-
hui*).[98]

The push for stock systems by nongovernmental enterprises was also
driven by the increasing complexity of enterprise management. The rou-
tinization of management, the increasing need to hold on to skilled personnel, and the desire to clarify ownership and authority relations within
the enterprise and between the enterprises and the supervisory agency
were all important motivating factors in the move to stock systems within
nongovernmental enterprises. In a 1993 survey of 271 enterprises in the
BEZ, enterprises most often listed "clarify property rights," "increase internal cohesiveness," and "perfect enterprise management" as the three
main goals of stock reform. Only 8.9 percent of respondents believed
that the main goal of stock reform was "accumulating development capital."[99]

One observer of high-tech enterprises in Beijing argued that managers
at Founders and Legend pushed for their companies to be listed on the
Hang Seng index in Hong Kong not simply to raise capital, but also to protect their own jobs.[100] During the late 1980s tension over the relationship
between Founders and Beijing University emerged between the two partners. Like many supervisory agencies, Beijing University believed that
Founders would have never been successful without the money it borrowed
from the university. Entrepreneurs acknowledged that the money had
been essential to their success, but they viewed it as a one-time loan, not a
continuing investment. Once Founders paid the loan back, it wanted to be
able to go its own way. The university wanted the enterprise to continue re-
mitting a portion of its profits even after the initial amount was paid back
and removed three managers who had tried to increase Founders' auton-

97. "95–96, Beijingshi Xinjishu Chanye Kaifa Shiyan Qu Fazhan yu Qushi Zonghe Baogao"
[Comprehensive report on trends and developments in the BEZ, 1995–1996], in *Beijingshi Xin-
jishu Chanye Kaifa Shiyan Qu Yanjiu Baogao: 1996* [Research report on the Beijing Experimental
Zone: 1996] (Beijing, 1997), 24–33.

98. Lianxiang Jituan [Legend Corporation], company prospectus, n.d.; interview, B27, 26
June 1997.

99. Out of five choices, clarifying property rights (*mingxi chanquan*) was listed most important by 53 percent of the respondents. Perfecting enterprise management was considered the
second most important goal by 35 percent, and collecting development capital (*jiju fazhan zijin*),
the fourth most important by 43 percent. Chanquan Zhidu Gaige Ketizu [Study group in property rights reform], "Beijingshi Yanqu Xinjishu Qiye Chanquan Gaige Diaocha yu Fenxi" [A report and analysis of property rights reform among BEZ enterprises] in *Beijingshi Xinjishu: 1995*
[Research report: 1995], 32.

100. Argument about use of stocks and the specifics of KeHai case are drawn from interview,
B10, 11 March 1997.

omy.[101] Managers in Beijing were also well aware of the case of Kehai, a non-governmental enterprise started with funds from the Chinese Academy of Sciences and the Haidian district government. After building Kehai up, the manager was removed because "personal relations went sour" with officials at CAS. Beijing entrepreneurs believed that with a listing on a stock market CAS administrators would have been more reluctant to remove an efficient manager for fear that the market would react negatively. At the very least, a shareholding system meant that managers would have a stake in a successful company they had helped build. The manager removed from Kehai left with nothing.

Issues of ownership and stock systems continue to be one of the main barriers to nongovernmental growth. In the early 1990s, the BEZ established a special agency to research the conversion of enterprise assets into stocks and the creation of "technology shares," but the policy environment has not kept up with changes on the ground.[102] Only two companies, Legend and Stone, have been able to restructure so far. In 2000, Legend officially ended its "nonstate" status, becoming a joint stock company, with the Institute of Computing Technology as one of the largest shareholders.[103] Under the plan, 35 percent equity share of Legend will be distributed to the company's employees. Of that amount, Legend's fifteen original founders will receive 35 percent, about 100 employees who have worked at Legend since 1988 will get a 20 percent share, and 45 percent will be distributed among "key players." Stone will let all employees voluntarily invest in an employee shareholding committee, which holds 51 percent of the shares of the newly registered Stone Investment. Stone Investment will eventually go public, allowing employees to sell their shares.[104]

GOVERNMENT SUPERVISION: POLITICAL
AND MARKET ACTIVITIES

Supervision was an extremely important complement to the financial and regulatory roles played by the local government. Government supervision was essential in encouraging talented individuals to enter the market in the first place; scientific personnel within the state S&T system had to be convinced to become risk takers. In Beijing, entrepreneurship was not concen-

101. By 1995, Founders was already on its third set of general managers, while most enterprises were still on their first. Stone was on its second but that was due to Wan Runnan's fleeing the country after Tiananmen.

102. Kou Zuopeng, "Beijing Xinjishu Chanye Kaifaqu de Gufenzhi Gaige" [Reforms of the stock system in the BEZ], *Zhongguo Keji Chanye* [Chinese S&T industries] 10 (1992): 10–12.

103. Hu Yanping, "Lianxiang/Jisuansuo—Ziji de Lihun Bieren de Piping" [Legend and the computer science institute—our divorce, other people's criticism] at http://www.sina.com.cn/news/review/2000-01-18/15840.shtm, accessed May 2000.

104. Cong Cao, "Zhongguancun: China's Silicon Valley," *The China Business Review* 28, 3 (May 2001): 38–41.

trated among those who had already finished a career in the state sector and were looking to take advantage of their political connections, but rather with those with benefits and status to lose. In 1996, one third of all *minying* personnel in Shanghai were retired cadres; only 5 percent of *minying* personnel in Beijing were retirees.[105] The rest either left the state sector before they reached retirement age or never entered it at all.

The local government's ability to encourage this type of entrepreneurship was extremely surprising given the view of Beijing as a city focused on politics, not business. Traditional regional stereotypes in China declare that southerners are more business savvy.[106] The creation of the institutional space for nongovernmental enterprises alone was not enough to encourage most scientists to set up their own enterprises and "jump into the sea" (*xia hai*). Once a scientist "jumped into the sea," they left the security of their state-sector job and the social welfare benefits it provided. Although most scientists knew they were not going to be guaranteed a life preserver once they entered the sea, they had to be reassured that at least a lifeguard patrolled the beach. With a range of formal and informal supervisory activities, the local government signaled to scientists that they would not be swimming completely alone.

Formal Supervision: The Beijing High-Technology Development Zone

Most formal supervision was institutionalized in the Beijing Experimental Zone. Although the first science park in the country was established in Shenzhen, the Beijing Experimental Zone quickly became the most successful and widely discussed high-tech zone in the Chinese media. Unique to the BEZ, the Beijing government created the zone in response to and after the rapid growth of *minying* enterprises; first the enterprises emerged, then the local government and the State Council created the zone. Support and research organizations were already closely tied and in close geographic proximity to production units. In other parts of the country, including Shanghai, Xi'an, and Guangzhou, the land for the zones was set aside, the zone created, and then enterprises moved into the area. By the end of 1994, the BEZ accounted for 25 percent of all income generated by zones, and 50 percent of all new products (out of fifty-two nationally recognized zones).[107] In 2000, enterprises in Zhongguancun invested 8 percent of their income in R&D, and technological developments contributed more than 50 percent to their profits; both figures are the highest for any enterprises in China.[108]

105. Interview, S8, 8 April 1996. See also, "Shanghai Minban" [Shanghai nongovernmental enterprises], *Zhongguo Keji Luntuan* 5 (1993): 3–5.

106. See, for example, the argument that "Beijing is a very political city, while Shanghai is an economic center." J. Bruce Jacobs, "Shanghai: An Alternative Center," in *China's Provinces in Reform: Class, Community, and Political Culture*, ed. David Goodman (New York: Routledge, 1997), 163.

107. "Beijing Xinjishu Chanye Kaifaqu Yao Jiduan Pandeng Xingaoshan" [Beijing experimental zone must continue climbing new heights].

108. See http://www.zgc.gov.cn, accessed September 2001.

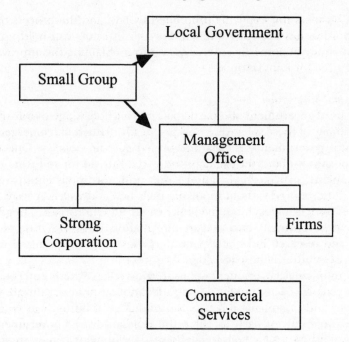

Figure 3.2. Management of the Beijing Experimental Zone

guidance to the zone, over involvement of the local government could have been a problem. In fact, the opposite occurred. By 1993, commentators argued that for the zone to move to the next level of development, it needed a management structure more able to coordinate between the various units involved in the zone. Moreover, the zone lacked the ability to direct investments in the state plan to *minying* enterprises, or the legislative power to initiate new policies that might meet the needs of growing *minying* enterprises.[120] As part of the Haidian district government, the management office could coordinate with the city government, but it had difficulties dealing with units in other administrative hierarchies.[121]

Still the local government has recently reiterated that a more hands-off approach is better suited to fostering innovation. In December 2000 the Beijing People's Congress approved "Regulations for the Zhongguancun Science and Technology Park." The regulations included articles on protecting enterprise assets and intellectual property rights, provided preferential policies for technological enterprises, and established an investment fund. Individuals and organizations were now permitted to undertake any

120. Lin Chen, "Beijing Shi Xinjishu Chanye Kaifaqu Mianlinde Kunnan, Wenti, he Zhengci" [Problems, difficulties, and policies faced by the Beijing experimental zone], *Keji Baodao* [Science and technology report] 10 (1992): 8–14.
121. Chen Zaofang, "Yingxiang Woguo Gao Xin Jishu Chanye Kaifaqu Fazhan de Yinsu Yanjiu" [Research into variables affecting the development of Chinese new and high-technology parks], *Keji Yu Fazhan* [S&T and development] (1992): 11–28.

type of business not explicitly prohibited by law, and the process of business approval has been streamlined. Foreign investors can hold up to 25 percent equity in domestic enterprises without obtaining the approval usually required for joint ventures.

Informal Supervision

The local government also participated in a whole range of informal activities; many of these roles were not specifically defined, but emerged from how local government officials interpreted specific policies. One report spoke positively about the Haidian Industrial Bureau for reducing capital requirements to make registration easier and not fining enterprises too hastily if they moved without reporting their new address right away. These activities were not illegal, but made life easier for enterprises.[122] In general, the local government encouraged information flows and fostered economic and research links between enterprises. It also legitimized *minying* enterprises in the Chinese ideological context.

The municipal government sponsored meetings where entrepreneurs could exchange ideas, publicizing new technological or organizational innovations and organizing job fairs for graduates of business or computer science programs. As early as 1984, the Haidian district government and CAS established a new technology joint development center to act as a bridge between enterprises and research units. This center was to be the node of a larger network that "not only affects technology enterprises, but the entire district."[123] In 1987 the city government established the Beijing Leading Small Group on Science and Technology (Beijing Shi Keji Lingdao Xiaozhu). This small group, under the direction of future mayor Chen Xitong, fostered meetings with more than one hundred S&T personnel and district officials in an attempt to create a network of enterprises in the city.[124] These meetings have continued throughout the reforms, and the Beijing local press has been filled with interviews with nongovernmental entrepreneurs discussing what else needs to be done to encourage the city's technological development. Local entrepreneurs used meetings at the beginning of 2001 to ask whether Zhongguancun can become the next Silicon Valley, to discuss the challenges facing nongovernmental enterprises after China's entry into the WTO, and to push for further clarification in property rights and for the ability to list on domestic and foreign markets.[125]

---

122. Wang Jianhua, "Sui wu zhu guan, dan you houdun" [Without supervision but still with support], *Zhongguo Keji Chanye* 8 (1998): 12–14.

123. "Kaichuang Zhongguoshi de Guigu de Tansuo" [Exploration of building a Chinese-style Silicon Valley], *Beijing Ribao*, 11 September 1984.

124. "Dongyuan Wange Keji Renyuan Zhiyuan Jingjiao Jianyi" [Mobilizing the suggestions of S&T personnel for Beijing's suburbs], *Keji Ribao*, 20 May 1987.

125. Mark O'Neill, "Silicon Valley Dream Remote from Reality," *South China Morning Post*, 21 February 2001.

There was also a range of actions unique to the Chinese ideological context. After the first *minying* enterprises emerged in 1980, articles appeared claiming these enterprises misused state-owned resources and that Chen Chunxian engaged in "dishonest practices" (*waimen xiedao*).[126] Statements in support of Chen in particular and science and technology personnel in general by central leaders like Fang Yi, Hu Qili, and Hu Yaobang insulated *minying* enterprises from further political attacks. *Minying* enterprises, like other parts of the economy, did not escape from the political fallout following the June 4, 1989, Tiananmen Square crackdown. Immediately after the massacre some district officials criticized *minying* enterprises as another form of private property. The founder of Stone, Wan Runnan, was forced to flee the country for his support of the students, and Stone made a public self-criticism in July 1989.[127] But political pressure on nongovernmental enterprises was short lived, and the local government moved quickly to assure high-tech entrepreneurs that its main concern was continued development. On September 23, Vice Mayor Chen Yucheng announced that the policy toward *minying* enterprises would not change; Wan Runnan's actions were his own and did not represent all *minying* enterprises. Chen stated that *minying* enterprises would continue to play a primary role in Beijing's development.[128] By 1990, the city government was ready to celebrate the ten-year anniversary of the birth of the *minying*, naming fifty-two enterprises as models for the rest of Beijing.

Tiananmen was the most prominent example of the political debate surrounding *minying* enterprises that continued well into the early 1990s. Deng's inspection tour of the South in early 1992 would be another boost to *minying* development, and technology officials popularized Deng Xiaoping's thoughts about the nature of nongovernmental enterprises. During the trip, Deng visited a number of high-tech enterprises, including several *minying* enterprises. At one, referring to the ongoing debate over whether *minying* were really part of the capitalist or socialist system, Deng reportedly exclaimed: "I do not want to continue this argument. From what I see here, nongovernmental are socialist."[129]

An equally important political activity occurred far away from the spotlight following Deng on his travels, and it occurred on a daily basis at the city and district level. During the middle to late 1980s, the proliferation of enterprises on Zhongguancun with no real technological capabilities and a tendency to cheat customers led to the witticism that the street was no

126. Beijing Nongovernmental S&T Association, *Beijing Minban: 1980–1990* [A chronicle].

127. For the text of the criticism see "Women Sitong de Renwu" [Stone's responsibility], *Zhongguo Minban Keji Shiyejia Xiehui Huixun*, 29 July 1989.

128. "Dui Minban Keji Shiye de Zhengce bu Bian" [The policy toward nongovernmental technology industries will not change], *Zhongguo Minban Keji Shiyejia Xiehui Huixun*, 9 November 1989.

129. The question is whether "*minban xing she xing zi*" [Are they called socialist or capitalist?]. Quoted in *1997 Niandu Beijing* [1997 overview].

longer "Electronics Avenue" (*dianzi yi tiao*), but "Crook Avenue" (*pianzi yi tiao*). Local officials explained that those doing the cheating were not *minying* but *getihu* (individual) enterprises. *Minying* entrepreneurs could not be capable of such dishonest practices since they were highly educated and of a better quality person (*suzhi hen gao*).

In one report, a journalist noted that people in Haidian district feared that high-tech industrialization and rapid growth would change the quality of their neighborhood. "Once computer companies take over," they asked, "how can we find a place to shop? We'll just be able to buy computers not tofu." The report assured these people that local leaders had been thinking about this problem, and that all would be made better off by the success of enterprises like Stone and Legend. According to the report, "If high-technology companies profit, then others also do well."[130] Statements like this legitimized *minying* enterprises both within the party ideology and the larger society.

Enterprise Networks

Though local governments adopted a wide range of social and political activities, expanding the role governments traditionally play in emerging markets, high-tech entrepreneurs also created their own market-supporting institutions. At least ten pricing periodicals were published by nongovernmental enterprises and entrepreneurs founded informal "Boss Clubs" to exchange information.[131] In cases where local government actions did not fit or conflicted with the interests of economic actors, entrepreneurs took advantage of the limited opportunities to create institutions semi-independent of government control. Local entrepreneurs established their own industrial association, Taisun Industrial Community (Taishan Chanye Jituan), approved by the All China Federation of Industry and Commerce. Founded in 1994 and made up of representatives of fifteen large nongovernmental enterprises in Beijing, including Stone, Legend, Kehai, and Jinghai, the community was organized like a *minying* enterprise. Taisun was managed by the "four self principles," and property collectively owned. The official goals of Taisun included "exchanging ideas and making contacts with enterprises in China and abroad, developing new industries in cooperation with members of the community, and studying the theory and experiences of modern enterprises so as to improve enterprise quality."[132]

In fact, Taisun evolved out of two sets of frustrations that *minying* entrepreneurs had with government officials. First, managers at the largest en-

130. "'Da Xiong Mao' Mianxiang Shijie" [A "giant panda" faces the world], *Beijing Ribao*, 5 March 1989.

131. Wang and Wang, "An Analysis of New Technology Agglomeration in Beijing," 694. Some are now published daily on the Web.

132. "Regulations for Taisun Industrial Community," Taishan Chanye Jituan [Taisun industrial community] prospectus, n.d., 5.

terprises felt the official nongovernmental technology entrepreneurs association concentrated on the needs of start-up and small-scale enterprises, ignoring the needs of the larger companies. As a result, most viewed the association's main goal as raising income through membership fees.[133] Second, entrepreneurs wanted to create an organization that addressed the central government's lack of concern for promoting the nonstate sector. According to one Taisun manager, central government officials did not truly understand the role of the modern enterprise in economic growth: "Enterprises should be the base of social and economic development, but the government has no goals for them except taxation."[134] Creating the alliance, the official continued, would not only allow high-tech entrepreneurs a chance to exchange ideas, pool investment, and provide other mutual help but also to negotiate with the government about the needs of high-technology entrepreneurs. The manager acknowledged that the current political situation made this unlikely, but in the future business people would be able to influence government policy.

Local entrepreneurs have also created a network of institutions that produce research reports and policy proposals reflective of the needs and interests of the *minying* sector. The Great Wall Enterprise Strategy Research Institute (Chang Cheng Qiye Zhanlue Yanjiusuo) was a nongovernmental enterprise that provided consulting advice to enterprises undergoing property rights reform and produced research reports on *minying* development.[135] Staffed by former research scientists with tight links to *minying* entrepreneurs and local government officials, Great Wall was emblematic of how the local government and high-tech entrepreneurs have worked together to create and reproduce a common vision of how the nongovernmental sector should be organized. Great Wall's 1995 report on the "second breakthrough" was funded by the Beijing Science and Technology Commission and written in cooperation with the research arm of the BEZ.[136] Overlapping networks and cooperative research ensured that local political and economic actors could adjust to each other's interests and create shared objectives for the sector.

CONCLUSION

National-level reforms of the science and technology system decentralized planning authority and reduced central government funding, shifting much of the responsibility of developing technology industries to local gov-

---

133. Interview, B30, 3 July 1997.
134. Interview, B28, 27 June 1997.
135. Interview, B29, 28 June 1997.
136. The report was a project of the Beijing Science Commission's Sociology of Science Plan. *Beijingshi Yanqu Gaoxin Jishu Qiye Erci Chuangye Zhanlue Yanjiu Baogao* [Research report on the BEZ high-technology enterprises' second breakthrough strategy] (Beijing, 1995).

ernments. Through reforms of the national S&T system, the central government created the institutional space for the birth of *minying* enterprises. Local authorities played a major role in interpreting how central government directives about technological reform should be implemented, thus creating new markets supportive of specific types of high-tech enterprises.

The range of policies available to municipal officials has not differed significantly throughout the country; all local governments have needed to secure capital and technology for new enterprises; all have used the development of high-technology zones and science funding to achieve these goals. The difference is that in Beijing local officials have promoted a definition of *minying* that revolved around independent, technologically competitive enterprises. In order to support and institutionalize this model as broadly as possible, authorities have been more open to experimentation in property rights, rarely intervened in the internal management of enterprises, and directed the majority of funding to either the general science budget or specific information-based technologies. Consequently, individual entrepreneurs were more successful in introducing and reproducing governance structures supportive of technological innovation.

Beijing's institutional and social structures made adopting this more nongovernmental model easier. The large concentration of S&T resources alone made science-based innovation and development more likely. In addition, as we will see in the next two chapters, such a high degree of agreement between local and central officials over the direction of local development policy was rare. Much of Beijing's early success may be the result of two conditions. On the one hand, with the central government's backing, the local government could aggressively support and nurture the high-technology sector. On the other, the type of support the local government could provide was severely limited; the local government had limited ties to and influence over the main actors in technological development. Under these conditions, the government's role was limited to what *minying* entrepreneurs needed most: the protection of enterprise autonomy, the facilitation of information flows, and the general improvement of Beijing's social and economic conditions.

Although this chapter has focused heavily on the role of government policy, this has not been a one-way process from local officials to high-tech entrepreneurs. Instead, local economic actors have been successful because they have been able to create, under extremely difficult conditions, enterprises that are more responsive to market signals. In Shanghai, Xi'an, and Guangzhou entrepreneurs have been much less successful in securing access to capital, technology, and skilled personnel while maintaining their enterprises' autonomy. Perhaps most important, entrepreneurs in these three cities, unlike in Beijing, have not yet been able to frame their interests in ways that make the local government a partner in development.

Shanghai: Small Enterprises
in a Big Enterprise Town

At the most basic level, Shanghai did not significantly trail Beijing in either the quantity or quality of science and technology resources. In 1994 Shanghai had 1,473 research and development organizations of all types compared to Beijing's 1,085, and Fudan, Jiaotong, and Shanghai Science and Technology universities were all at or near the top of the educational system.[1] In the same year, of total national science and technology (S&T) expenditures, RMB 10.24 billion was spent in Beijing; Shanghai followed not far behind with RMB 9 billion.[2] In addition, Shanghai fell under the same set of national policies to support the growth of technological enterprises, and municipal officials declared that the information industries were to become a "pillar industry" (*zhizhu chanye*) of the local economy.

The results of nongovernmental technology enterprise development, however, differed. Large, state-owned enterprises were at the center of Shanghai's technological development, and the information technology (IT) sectors that have been most successful locally are the ones most amenable to planning and coordination. Beijing has led the country in software and internet technologies, Shanghai in the manufacturing of IT hardware, especially integrated circuits. During the first decade of reform, Shanghai trailed behind Beijing in the number of nongovernmental enterprises. In 1990, Shanghai only had 607 registered nongovernmental enterprises compared to 2,100 in Beijing. From 1988 to 1994, Beijing nongovernmental enterprises received 32.5 percent of all awards given to National Technology Entrepreneurs, more than in any other part of the country.[3] Total income of all *minying* enterprises in Shanghai in 1993 equaled one tenth of those in Beijing's Haidian district; exports were a little more than half; and only 16 percent of products were designated high

---

1. This includes independent research institutions, research labs attached to state-owned enterprises, and institutes of higher learning. *Zhongguo Keji Tongji Nianjian: 1995* [China statistical yearbook on science and technology: 1995] (Beijing: Zhongguo Tongji Chubanshe, 1995), 4.

2. Xinhua News Service, 29 July 1995, in *Foreign Broadcast Information Service-China* (hereafter FBIS-CHI), 39, 95–146.

3. "Beijing Minying Keji de Chenggong Moshi" [Patterns of Beijing nongovernmental enterprise's success], *Zhongguo Keji Chanye Yuekan* [Chinese technology industry monthly] 9 (1995): 27–28.

tech, as opposed to 20 percent in Beijing.[4] The gap in the numbers of non-governmental enterprises has gradually closed. After Deng Xiaoping's inspection tour, the number of nongovernmental enterprises took off; by the mid 1990s, there were over 7,000 *minying* enterprises in Shanghai. But technological development in Shanghai still is not driven by what happens in these enterprises; state-owned enterprises remain the engine of growth. The six largest state-owned enterprise groups generated 87 percent of total output value in the IT sector in 2000.[5]

From the beginning of the reform process in Shanghai, it was clear that *minying* enterprises were not at the center of the local government's development plan. A book published by the research office of the Shanghai Communist party Commission addressed the fear that promoting high-tech companies and cooperating with foreign multinationals (MNCs) in developing advanced technologies would undermine state-owned enterprises (SOEs). The authors argued that this was exactly the opposite of what would happen: "All the famous high-tech industries are under the city's control and they will greatly strengthen state-owned enterprises' position in the national economy."[6] High-technology industries were to bolster the position of traditional industries; *minying* enterprises were to provide a good example to state-owned enterprises about how internal management could be improved, but they were not to displace those enterprises from the commanding heights of the local economy.

The Shanghai government did not question that *minying* enterprises could be more flexible and more responsive to market competition than SOEs. Planning and research documents in Shanghai also spoke of the vibrancy of the nongovernmental sector, of enterprises responsible for their own success and failure. But there was in these same documents a recurring suspicion about whether Shanghai could and should rely on small scale, autonomous enterprises to achieve its economic goals. This suspicion revealed itself in the ways Shanghai implemented the funding, regulatory, and supervisory policies that affected *minying* enterprises. Shanghai concentrated most of its R&D budget within large state-owned industries. While Beijing often acted early to support *minying* enterprises, Shanghai mainly ignored the earliest *minying* entrepreneurs. And while local officials in Beijing signaled official interest and concern for the sector, creating an environment more supportive of entrepreneurship, Shanghai neither encouraged scientists to "jump into the sea" (*xia hai*) and start their own

4. Zhu Jianjiang, "Shanghai Minban Keji Qiye de Yunxing Jizhi ji Guanli" [The operation and management of Shanghai nongovernmental science enterprises], *Zhongguo Keji Luntan* [China S&T forum] 5 (1993): 42.

5. "Vice Mayor: Shanghai Leads PRC IT Industry," *Xinhua*, 9 May 2000, in FBIS-CHI, 10 May 2000.

6. Lin Qizhang, *Shanghai Fazhan Yanjiu: Gaoxinjishu ji Chanye Lun* [Shanghai development research: High-technology and industries] (Shanghai: Shanghai Yundong Chubanshe, 1995), 107.

companies nor acted as lifeguard, supporting those who did. Shanghai's inability (or unwillingness) to support independent entrepreneurship and its continued focus on SOEs fragmented technological networks, isolating entrepreneurs from the local government and each other.

At the beginning of the reform period, Shanghai was severely constrained by its important role in the national economy. National leaders did not allow experimentation that threatened Shanghai's financial remittances to the center, and the large concentration of SOEs in the region meant that public-sector managers could lobby against policies supportive of nongovernmental enterprises. However the local context has gradually changed. Some district officials began paying closer attention to nongovernmental enterprise development in the early 1990s, and the municipal government has recently made a more concerted effort to foster small enterprise growth. Still, for the most part, the Shanghai government continues to reproduce a development strategy centered on large enterprises, horizontal coordination, and state intervention.

THE SHANGHAI DEVELOPMENT PATH

Shanghai has always attracted intense central government attention.[7] Despite the difficulties of clearly demarcating the battles between the center and the city that raged throughout the reforms, it is clear that for most of the period Shanghai was severely restricted in the range of development opportunities open to it.[8] Until the mid 1980s the comparative advantages of the city—commerce, light industry, and international finance—were subsumed in and sacrificed to larger national development projects.

After taking power in 1949, the Chinese Communist party, suspicious of the city's ties to the world and concerned about unequal regional development, cut the city off from international trade, restricted private commerce, and focused on heavy industry. Before the "Five Anti's" campaign, a 1952 movement directed at alleged corruption and crime in the private sector, private manufacturing accounted for 80 percent of industrial ca-

---

7. In fact, the patterns of Shanghai's more interventionist policies were set before 1949. During the 1920s and 1930s, the Nationalist government was more intrusive in Shanghai than in other parts of the country, using the newly established court system to enforce licensing on a wide range of businesses. Beijing in the same period was characterized by the increasing independence of social actors. See William Kirby, "China Unincorporated: Company Law and Business Enterprises in Twentieth-Century China," *Journal of Asian Studies* 54, 1 (February 1995): 43–63; Christian Henriot, *Shanghai, 1927–1937: Municipal Power, Locality, and Modernization* (Berkeley: University of California Press, 1993); and David Strand, *Rickshaw Beijing: City, People, and Politics in the 1920s* (Berkeley: University of California Press, 1989).

8. David Goodman, "The Shanghai Connection: Shanghai's Role in National Policies during the 1970s," in *Shanghai: Revolution and Development in an Asian Metropolis*, ed. Christopher Howe (New York: Cambridge University Press, 1981), 128.

pacity in Shanghai.[9] Despite Mao's early assurances that the transition to socialism would be gradual, the private sector was completely socialized by 1956; private industry, according to the First Five-Year Plan, was to be "reformed, controlled, and restricted."[10] Private industry disappeared as entrepreneurs either fled to Hong Kong or were absorbed by collective enterprises. The ratio of light to heavy industry in the city fell from 71:29 in 1957 to 49:51 in 1978.[11]

Central development plans resulted not only in the loss of Shanghai's comparative advantage in light industry, but also the loss of economic autonomy. Between 1949 and 1983, about 87 percent of the Shanghai city government's revenue was remitted to the center.[12] Since the central government and not the local authorities determined how the revenues that were collected would be spent, political scientist Lin Zhimin describes Shanghai as "revenue rich but resource poor."[13] The resource outflow resulted in the chronic under funding of the local infrastructure. The growth of the state sector and SOEs did little to increase Shanghai's financial capabilities. Ninety percent of all local enterprise earnings were controlled by central planners in Beijing.[14]

With the start of the reforms in 1978, common sense suggested that the central government look to Shanghai to be the country's most important supplier of electronics products. This attitude reflected the traditional importance of Shanghai in China's overall development and the strength of its R&D base. In addition, Shanghai already played a leading role in the consumer electronics sector. In 1981, for example, Shanghai produced 22 percent of all televisions and 37 percent of all tape recorders in China.[15] In the long term, the central government could expect Shanghai to take advantage of its high concentration of research institutes and skilled manpower to help build a computer industry.

But even with these comparative advantages, the central government

9. Christopher Howe, "Industrialization under Conditions of Long-Term Population Strategy: Shanghai's Achievements and Prospects," in *Shanghai: Revolution and Development*, 166.

10. Quoted ibid., 168.

11. Zhimin Lin, "Shanghai's Big Turnaround since 1985: Leadership, Reform Strategy, and Resource Mobilization," in *Provincial Strategies of Economic Reform in Post-Mao China: Leadership, Politics, and Implementation*, ed. Peter Cheung, Jae Ho Chung, and Zhimin Lin (Armonk, N.Y.: M.E. Sharpe, 1998), 51.

12. J. Bruce Jacobs and Lijian Hong, "Shanghai and the Lower Yangzi Valley," in *China Deconstructs: Politics, Trade, and Regionalism*, ed. David S.G. Goodman and Gerald Segal (New York: Routledge, 1994).

13. Zhimin Lin, "Reform and Shanghai: Changing Central–Local Fiscal Relations," in *Changing Central–Local Relations in China: Reform and State Capacity*, ed. Jia Hao and Zhimin Lin (Boulder, Colo.: Westview Press, 1994), 239–60.

14. Shahid Yusuf and Weiping Wu, *The Dynamics of Growth in Three Chinese Cities* (New York: Oxford University Press, 1997), 48.

15. Detlef Rehn, "Organizational Reforms and Technology Change in the Electronics Industry: The Case of Shanghai," in *Science and Technology in Post-Mao China*, ed. Denis Fred Simon and Merle Goldstein (Cambridge, Mass.: Harvard University Press, 1989), 144.

was ambivalent about Shanghai's playing the vanguard role in high-technology development. The central authorities, as they have since 1949, wanted to get as much out of Shanghai as possible; the center saw Shanghai as a source of basic electronics products, but not as a catalyst to overall economic development. Modernizing Shanghai's industrial base would mean either a significant financial commitment from the center or allowing Shanghai the freedom to go its own way. Local financial resources were limited, and taking some factories off line for restructuring meant a reduction in revenues remitted by Shanghai to the center. As one local government official argued, "Shanghai is so important to the national economy that the central government was less likely to allow experimentation that might threaten its revenues. Failure in Shanghai would affect the entire country."[16]

By the mid 1980s, leaders in Beijing had begun to change their minds about Shanghai's role in the national economy. Central leaders feared that a low growth rate of industrial production and slow progress in science and technology would seriously affect the local economy and thus harm the entire modernization program.[17] While the southern provinces, especially Guangdong, seemed to be sprinting through the door that the reforms had opened, Shanghai was falling behind. In 1978, Shanghai topped the list of all provinces and regions in contributions to the national income; by 1986 Shanghai had fallen to number six, in 1990 to number ten.[18] Guangdong replaced Shanghai in 1986 as the primary exporting province.[19]

By February 1985 a new development plan had been accepted for Shanghai. Shanghai was to become a multifunctional center, regaining its national lead in commerce, trade, and science and technology, updating existing industries and opening to the world economy. Local control over fiscal revenues increased, and the city was allowed to raise funds on international markets. Shanghai's development strategy was reinterpreted again between 1988 and 1991, focusing on the development of the new Pudong district, the renovation of aging infrastructure, and the strengthening of the city's role as a foreign trade, finance, and service center. In addition, the city adopted a two-pronged attack on the problem of upgrading existing industrial technology: slow, incremental changes within SOEs, and more rapid development in new pillar industries including information technologies.

16. Interview, S14, 18 April 1996.
17. Victor Mok, "Industrial Development," in *Shanghai: Transformation and Modernization under China's Open Policy,* ed. Y. M. Yeung and Sung Yun-wing (Hong Kong: Chinese University Press, 1996), 199–224.
18. Peter Cheung, "The Political Context of Shanghai's Economic Development," in *Shanghai: Transformation and Modernization,* 55.
19. *Zhongguo Tongji Nianjian 1995* [Statistical yearbook of China 1995] (Beijing: Zhongguo Tongji Chubanshe, 1995).

Policy Traditions

The growing difficulties of state-owned enterprises that emerged later in the reform period were a heavier burden in Shanghai than in any other city discussed in this book, except perhaps Xi'an. While the state sector contributed between 20 and 25 percent of industrial output in other coastal areas like Jiangsu and Guangdong, SOEs made up close to 40 percent of output in Shanghai. In 1994 the state sector still employed 72 percent of the urban workforce, 81 percent of all first jobs were allocated through the state, and all types of private industry made up only 2.8 percent of local industry.[20] Subsidies in the city for SOEs increased six times from 1987 to 1993 to RMB 3.35 billion.[21] In addition to the burden the public sector placed on local finances, the dominance of SOEs made it harder for other sectors to develop. During the early 1990s the tertiary sector was 1.4 times larger in Guangzhou than in Shanghai.[22] As one manager at a nongovernmental enterprise noted, "SOEs are huge in Shanghai—they are like trees blocking out the sun."[23] Finally, the dominance of SOEs shaped the organization of new sectors. Even when nongovernmental enterprises were formed they tended to re-create the structures of SOEs. As the sociologist Neil Fligstein argues, "At the beginning of a new market, the largest firms are the most likely to be able to create a conception of control and persuade others to go along."[24]

The methods of central planning were internalized and reproduced by local leaders. Even after the decentralization that occurred during the reforms, Shanghai's development strategy was one of "high input, high risk, and a high level of reliance on government guidance."[25] Constant central attention meant that the organs of local government were more interventionist than they were elsewhere. As one central government official explained: "In Shanghai, the local government dominates and it knows how to take advantage of the old planning system. Even branches of central state ministries must first answer to the local government."[26] The journalist Pamela Yatsko describes a city determined "to use state-planning methods to advance its manufacturing prospects in China's budding market economy."[27]

Reforms of the central state S&T system have often resulted not in the

20. Wong Siu-lun, "The Entrepreneurial Spirit: Shanghai and Hong Kong Compared," in *Shanghai: Transformation and Modernization,* 39.

21. Cheung, "The Political Context of Shanghai's Economic Development," 56.

22. Lee Wilson and Brian Hook, "Human Resources," in *Shanghai and the Yangtze Delta: A City Reborn,* ed. Brian Hook (New York: Oxford University Press, 1998), 123.

23. Interview, S35, 17 June 1998.

24. Neil Fligstein, "Markets as Politics: A Political-Cultural Approach to Market Institutions," *American Sociological Review* 61 (August 1996): 664.

25. Lin, "Shanghai's Big Turnaround," 67.

26. Interview, B25, 30 May 1997.

27. Pamela Yatsko, *New Shanghai: The Rocky Rebirth of China's Legendary City* (New York: John Wiley, 2001), 252.

devolution of economic power to independent enterprise level units, but in the strengthening of the administrative controls of the municipality. For example, in response to the March 1985 Decision on Reforming S&T Management, the Ministry of Electronics (MEI) institutionalized a system of branch management to overcome the vertical barriers that separated the various units involved in the electronics industry.[28] All enterprises under the direct control of the MEI were to be placed under local supervision in Shanghai. After being removed from MEI control, these enterprises were linked to local government institutions to form branch corporations that would integrate research, production, sales, and service institutions.[29] A policy decision that helped shore up the growing *minying* industry in Beijing was used to increase local government efficacy in Shanghai.

This focus on economic planning and government guidance was reflected in the balance of power among local government institutions. Ministries traditionally active in the state plan and sympathetic to the plight of SOEs, such as the planning and economic commissions (*jiwei* and *jingwei*), dominated bureaucratic politics. Members of the Science and Technology Commission rarely worked their way up the ladder of city politics, and none was appointed mayor or vice mayor in the 1980s or 1990s. In fact, one interviewee reported that members of the Science and Technology Commission were the least respected in city government, seen to have little real economic knowledge and often derided as "bookworms" (*shudaizi*).[30]

Social Base of Entrepreneurship

Shanghai's development history severely affected the range of possible partners the local government could have cooperated with if it had decided to focus on supporting *minying* enterprise development. Private entrepreneurs fled after the 1949 revolution, damaging the commercial capabilities of Shanghai. Skepticism about the attractiveness of entrepreneurial careers remained in Shanghai. Most interviewees remarked that talented individuals would rather become managers of larger enterprises than start their own; a 1988 survey found that only 1 percent of respondents said they would prefer to be self-employed compared to 40 percent in Hong Kong.[31] The outflow of entrepreneurial spirit was paralleled by a relocation of S&T talent to interior areas. The central government adopted a policy of sending skilled technicians to the countryside to help build rural industries. During the thirty years between 1950 and 1980, 1.4 million

28. Dianze Gongye Bu, "Guanyu Tuijin Dianzi Gongye Jingji Guanli Tizhi Gaige de Baogao" [Report on the implementation of the reforms of the management system of the electronics industry], *Zhongguo Dianzi Bao* [China electronics daily], 21 March 1986.

29. Rehn, "Organizational Reforms and Technology Change," 141.

30. Interview, S29, 28 October 1996.

31. Wong Siu-lun describes "enterprising Shanghainese as individuals, but not an enterprising Shanghai as a community." See "The Entrepreneurial Spirit," 34.

people left Shanghai, 30 percent of them technicians.[32] While these policies may have helped build industries in remote areas, they severely
drained Shanghai's reserve of S&T talent.

The lack of a large concentration of skilled personnel in Shanghai also
influenced the political prospects of *minying* enterprises. A large pool of
local scientists may have been able to develop personal connections to local
officials and lobby for better support of nongovernmental enterprises. As it
was, unlike their counterparts in Beijing, Shanghai entrepreneurs did not
have well-developed connections to highly placed officials. They did not
have crucial allies in the central and local government able to insulate technology enterprises from harsh ideological criticisms like CAS did for enterprises on Zhongguancun. Conversely managers in local SOEs did have extensive ties to local officials, and they were likely to use them to lobby for
their own interests and against those of *minying* enterprises.

The Lure of the Large: Shanghai's Model of Innovation

Influenced by the success of technology firms in Silicon Valley, Beijing
consistently promoted a definition of *minying* that highlighted the nongovernmental nature of these enterprises. Moreover, decision-makers in
the municipality framed technological development as moving from the
local economy out to the national and, eventually, international
economies. Growth was viewed as a gradual process, emerging first at the
enterprise, and then proceeding to the district, municipal, national, and
international levels. Shanghai's development formula drew from different
sources and followed a different logic. Local planners spoke more of the
examples of Japan and Korea than northern California or Route 128.
Shanghai's model was of state-led development and large industrial groups.
A book published in 1996 for local science commission cadres, for example, favorably described how Japan after the war had first looked for advanced technologies from abroad and then gradually developed its own innovative ability through large conglomerates.[33]

Moreover, Shanghai, from the beginning stages of reform, focused on its
role in the national as well as regional and international economies. As a result of the need to compete in so many different markets simultaneously, officials argued that bigger enterprises were more likely to be competitive,
and technology must be developed both in new sectors and as a way to regenerate traditional industries. Discussions of technological development in
the newspapers began with descriptions of the decrepit state of Shanghai's
industrial structure; 70 percent of equipment dated back to 1950s and
1960s, while only 5 to 7 percent of equipment was equal to an international

32. Wilson and Hook, "Human Resources," 126.

33. Yang Fujia, ed., *Xiandai Keji yu Shanghai* [Modern science and technology and Shanghai]
(Shanghai: Shanghai Kexue Puji Chubanshe, 1996), 367.

level of the early 1980s.[34] Since *minying* enterprises tended to be small, they were little help in attaining the city's development goals. When local leaders did pay attention to nongovernmental enterprises, it was to encourage them to grow as quickly as possible and to eventually form large conglomerates (*jituan*). Even when they did achieve a significant scale of production, *minying* enterprises were viewed as a useful complement (*buchang*) to the state economy; they were not to be allowed to displace SOEs.

Throughout the reforms, local officials made public proclamations about the importance of technology to Shanghai's growth generally, and about the need to develop information technologies in particular. In the words of Vice Mayor Jiang Yiren, "High technology is proof of a local economy's competitive power." But developing high-technology industries, the vice mayor continued, cannot be separated from other concerns; "Science and technology development and industrial adjustment must proceed together."[35] Even academic researchers of technology development stressed that high-tech innovation needed to be linked to Shanghai's other development plans, arguing that technological innovation referred not only to high-technology enterprises, but also to the use of high-technology to reform traditional state-owned enterprises.[36]

Like innovation itself, Shanghai's growth could not be separated from the larger economic context. High-tech development was only one of the many possible paths to Shanghai's achieving its larger development goals. Moreover, Shanghai framed its economic goals in larger terms than just its own development.[37] Under Mayor Xu Kuangdi, Shanghai described its economic role as "one dragon head, three centers" (*yi ge longtou, san ge zhongxin*).[38] In this construction, Shanghai was to be both a regional economic center and an international economic, trade, and financial center.

This focus on the large was reflected in the debate about creating a mod-

---

34. Tai Shijun, "Guanyu Fazhan Shanghai Gao Xin Jishu Chanye de Zhanlue Yanjiu" [Research on developing Shanghai's new and high-technology industry strategy], *Shijie Kexue* [World science] 9 (1987): 28–31.

35. Jiang Yiren, "Jiakuai Gaoxinjishu Chanye Kaifa Qu Jianshe, Zujin Gaoxinjishu Chanye Fazhan" [Speed the construction of the HT industry development zone, promote HT industrial development], in *Guojia Gao Xin Jishu Chanye Kaifaqu Suozai Shi Shi Zhang Zuo Tanhui: Wenjian Zailiao Huibian* [Collected documents from the meeting of mayors from cities with national level new and high-technology industry development zones] (Beijing: Guojia Kewei Huaju Jihua Bangongshi, 1996).

36. Xiao Yuanzhen, "Pudong Xinqu Gaojishu Chuangye Zhanlue Gouxiang" [Framework for Pudong new district high-technology industrial strategy], *Keji yu Fazhan* [S&T and development] 4 (1992): 29.

37. Central leaders during visits to Shanghai have also discussed its technological development in the context of its regional and international role. See for example, *Shanghai Keji Ribao*, 10 May 1996.

38. The slogan was popular during 1992–94 and was part of a research project that was eventually published in three books edited by Cai Laixing, Xu Qiang, and Mayor Xu Kuangdi. The dragon head refers to Shanghai's leading role in the economic region that lies along the Changjiang River, since the river is seen to resemble a dragon that spreads from west to east across the country. I thank Wu Xiaogang for this information.

ern enterprise system. In discussing technological development, planning documents from Shanghai focus on the central role of the modern enterprise. But unlike in Beijing where a growing network of nongovernmental enterprises would be the basis of reform, Shanghai tried to reform the old system, not build a new one. Leaders like Huang Ju were clear that "the main part of scientific progress is the enterprise, especially those large state-owned enterprises that occupy a dominant position in the national economy."[39] Focusing on the large meant ignoring *minying* enterprises. Proponents of supporting nongovernmental growth complained that local planners were concerned only about large enterprises. As one commentator noted, the local government adopted a policy of "grasping the big, letting go of the small" (*zhua da, fang xiao*). This was understandable, according to the article, but the commentator continued that Shanghai should also "pay attention to small enterprises that would eventually become large enterprises."[40]

During the early and mid 1980s, Shanghai officials rarely spoke of nongovernmental enterprises as enterprises, as if doing so would put them on the same level as enterprises in the public sector. The rare early public pronouncements about the first *minying* enterprises in Shanghai referred to them as organizations (*jigou*), not enterprises (*qiye*).[41] By the early 1990s, when pronouncements about the enterprises became more common, they were almost always in the context of their small scale of production or informal management practices, characteristics viewed as weaknesses to be overcome.

Underlying these ideas about the organization of innovation was a view of the role of government that differed from the one that dominated in Beijing. In contrast to Beijing's efforts to protect the autonomy of nongovernmental enterprises from outside interference, the story of *minying* development in Shanghai was one of neglect. *Minying* enterprise did not fit traditional patterns of industrial policy and so fell outside the concern of local officials. The only response to *minying* development that Shanghai officials could come up with was to make them bigger. The problem with neglect was that it occurred in tandem with efforts to coordinate other parts of the economy. Beijing officials worked to create an institutional space for *minying* enterprises; Shanghai did little to create that space, and *minying* enterprises were increasingly squeezed by official efforts to coordinate other parts of the economy. As one former member of the local Sci-

39. Huang Ju, "Jiasu Keji Jinbu, Zujin Jingji Fazhan he Shehui Quanmian Jinbu" [Speed S&T progress, promote economic development and overall social progress], in *Shanghai Kexue Jishu Gongzuo Nianbao: 1995* [Shanghai yearbook of science and technology work: 1995] (Shanghai, 1996), 4.

40. Wang Pusheng, "Minban Keji Fazhan Zhong de Wenti Yu Jianyi" [Problems in and suggestions for developing nongovernmental S&T], *Minban Keji* [Nongovernmental S&T] 6 (1991): 8.

41. See for example, the 1989 Shanghai Shi Minying Jigou Guanli Tiaoli [Regulations of nongovernmental institutions].

ence and Technology Commission argued, "Officials in Shanghai believe that state control over all types of enterprises must remain strong, and managers within enterprises should also retain power over all types of decisions. Managers do not want to release hold over salaries or S&T personnel and so enterprises are managed to death (*guan si le*)."[42]

The success of nongovernmental enterprises in Beijing and other places began to influence officials in Shanghai. Beginning in the early 1990s, local officials, especially at the district level, investigated nongovernmental enterprise growth in Haidian district.[43] Consequently, they adopted a different vocabulary to describe a different set of policies. As Vice Mayor Liu Zhenyuan stated in 1993, "Our future responsibility is to support a system that supports *minying* enterprises."[44] Still, SOEs and large business groups remain at the center. Shanghai's Tenth Five-Year Plan (2001–2005) calls for information technology to become the city's preeminent sector and states that one of the main priorities is to create Chinese-owned companies able to compete with MNCs in domestic and foreign markets. In order to accomplish this goal, local resources will be funneled into six industrial groups.[45]

ROLE OF THE SHANGHAI GOVERNMENT: FINANCE

The shortage of start-up and venture capital that severely limited the rate of growth of *minying* enterprises in Beijing had a similar effect in Shanghai. The first *minying* enterprise emerged in 1984 as a technology consulting company, and those that followed raised funds in ways similar to other *minying* enterprises throughout the country.[46] In a 1985 survey of joint science and production units, 60 percent of the enterprises still had the original research or educational unit "as the core investing unit;" 25 percent were new enterprises that were set up by the original research institute and a production unit, probably a state-owned enterprise.[47] Fudan University professor Chen Suyang founded Forward with his own savings and some investment from friends; Fudan provided access to university-owned offices

42. Interview, S36, 19 June 1998.
43. "Putuo Qu Kewei Wei Minke Qiye Fazhan" [Putuo district science commission in support of nongovernmental enterprise development], *Zhongguo Minying Keji Yu Jingji* [China nongovernmental S&T and economy] 4 (1996): 42.
44. Quoted in Zhu Jianjiang, "Shanghai Minban Keji Qiye" [The operation and management of Shanghai], 43.
45. Stephen Harner, "Shanghai's New Five-Year Plan: The Pearl Starts to Shine," *China Online,* 18 December 2000.
46. Respondents in Shanghai tended to claim that the first *minying* emerged at the same time as in Beijing, around 1980. A 1988 news report states the first *minying* was founded in Shanghai in 1984. See "Shanghai Shi Chengli Minban Keji Shiyejia Xiehui" [Shanghai establishes nongovernmental industrialist association], *Keji Ribao* [Science and technology daily], 26 February 1988.
47. Gu Wenxing, "Shanghai Keji Tizhi Gaige de Wuxiang Cuoshi" [Measures for reforming Shanghai's S&T system], *Kexuexue yu Kexue Jishu Guanli* 4 (1985): 16–17.

and equipment, but no investment capital. Fuxing, another technology company with ties to Fudan, started as a small consulting company with registration capital of only RMB 100,000 raised by the founding members. A large number of nongovernmental enterprises registered as collective and were actually divisions of state-owned enterprises, and they received their initial investment and operating capital from those enterprises.

Bank lending was restricted because of the needs of the local government and the concerns of the banks. The 1985 Decision Concerning the Reform of the Science and Technology Management System recognized nongovernmentals at the national level and made Shanghai *minying* enterprises eligible for bank loans.[48] This did little for nongovernmental enterprises; the local government used bank lending to shore up foundering SOEs. From 1985 until 1993 city financial organs and banks issued loans for more than RMB 300 million to state-owned high-tech enterprises.[49] Still this was not enough to address the key limitation to *minying* enterprise growth: scarce capital. In a 1993 survey of Shanghai technology enterprises, 53 percent declared the biggest problem to continuing development was "the lack of adequate capital."[50]

While the 1993 Decision on Several Problems Facing the Enthusiastic Promotion of Nongovernmental Technology Enterprises made it officially easier for nongovernmental enterprises to approach banks for funding, changes in the banking system made it more difficult to find banks willing to make those loans. The reform of the banking system converted banks into commercial ventures; they no longer acted simply as the disbursement arm of the central plan and were increasingly concerned with returns on loans. Banks in Shanghai viewed *minying* enterprises as high-risk borrowers since few grew rapidly enough to make repayments, and some *minying* enterprises actually defaulted on early loans, making banks wary about issuing further loans.[51] In this context, not having local officials, either at the city or district level, willing to vouch for their creditworthiness meant most Shanghai nongovernmental entrepreneurs faced severe capital shortages. Enterprises like Forward that had extremely good connections with the local government managed to secure loans from local banks. By 1996, Forward had received short and long-term loans totaling RMB 68 million from the Industrial and Commercial Bank (Yangpu branch), Communications Bank (Jiading branch), and Bank of China (Shanghai branch).[52]

48. Tony Saich, *China's Science Policy in the 1980s* (Atlantic Highlands, N.J.: Humanities Press International, 1989), 161–64.

49. This figure is not broken down by enterprise type. Lin Huangu, "Shanghai Gaojiao Keji Chanye Xiankuang he Fazhan Duice de Yanjiu" [The current status and development policy for technology industries established by institutes of higher education], *Zhongguo Minying Keji Yu Jingji* [China nongovernmental S&T and economy] 2 (1995): 15.

50. "Shanghai Gaoxin Jishu Chanyehua de Chengyin he Wenti" [Accomplishments and difficulties in Shanghai's high-technology industrialization], *Kexuexue yu Kexue Jishu Guanli* 10 (1993): 21–23.

51. Lin Huangu, "Shanghai Gaojiao Keji" [The current status and development policy], 15.

52. Shanghai Forward Group Limited, *1996 Annual Report*, 17–18.

Funding from Joint Ventures

In addition to banks, the local government pursued other means of raising capital for technological development. In particular, Shanghai developed a strategy highly reliant on foreign corporations. The Shanghai Science and Technology Commission in 1989 surveyed managers of technology enterprises about how best to develop new electronics products. Of those surveyed, 40 percent believed that Shanghai had the best chance of developing competitive products by relying on foreign partners for new technologies and capital; 32 percent said they should cooperate with domestic research or educational organizations. Only 4 percent of respondents believed that new technologies could be developed solely within their own factories.[53]

In order to attract this capital, foreign enterprises in Shanghai proper were exempted from local income, land use, and housing taxes for the first three years after their establishment and were excused from paying housing subsidies for Chinese workers.[54] In 2000, the local government announced revised guidelines for foreign investors in high technology.[55] The new policy stipulated that the land use fee would gradually be repatriated for the first three years, foreign ventures would be able to pay reduced business and income taxes for five years, and foreign ventures with input in research and development that has increased by more than 10 percent from the year before would enjoy income tax cuts equal to 50 percent of the input.

The results of these preferential policies were that the scope of foreign investment in technological enterprises was larger in Shanghai than Beijing. By 1993, twenty-seven multinational corporations invested a total of $190 million, or an average of $7 million per venture. In Beijing seventy MNCs averaged an investment of $710,000 for a total of $49 million.[56] Shanghai's focus on integrated circuit manufacturing has reinforced concentration of foreign investments. Taiwan has invested $1.63 billion in a microchip foundry; IBM announced a project that includes a $300 million silicon chip assembly plant; and Japan's NEC contributed $500 million to a joint venture in chip fabrication, Shanghai Hua Hong NEC Electronics.

Linked to state-owned enterprises through joint venture agreements

53. Zhang Min, "90 Niandai Woguo Gao Jishu he Gao Jishu Chanye Fazhan Chuxi" [Chinese high-technology and technology industries development in the 1990s], *Keji Guanli Zixun* 11 (1991): 6–8.

54. Enterprises within the zones received even more preferential policies. See Shanghai Municipal People's Government, "Provisions of the Shanghai Municipality for the Encouragement of Foreign Investment," 23 October 1986.

55. "Shanghai to Launch Preferential Policy for Attracting Hi-Tech Funds from Abroad," *Xinhua* in FBIS-CHI, 28 December 2000.

56. The Beijing number may also include "three capital" enterprises, which often refers to enterprises owned by Hong Kong, Taiwan, or overseas Chinese. See "Beijing Shiyan Qu Yao Jiduan Pandeng Xin Gao Shan" [BEZ must continue climbing new heights], *Zhongguo Keji Luntan* 3 (1994): 43–47.

brokered by the Municipal Foreign Economic Relations and Trade Commission, these foreign firms were expected to make up deficits in Shanghai's investment capital, production ability, and managerial skills. Shanghai Belling was the most widely reported success story. Established in 1988 by the Shanghai no. 14 Radio Factory and Shanghai Bell Telephone Equipment, a local branch of the Belgian Bell Company, Belling produced 24.8 million chips in 1994, with profits of RMB 130 million.[57] The goal of creating enterprises like Shanghai Belling was to protect and strengthen the position of SOEs in the public sector and to raise the technological level of the local economy as a whole. Small, autonomous nongovernmental enterprises had a limited role to play.

Direct Funding: Science Budgets

As happened in Beijing, infrastructure projects and investment in the local science budget helped create a more competitive environment for nongovernmental enterprises. These types of investments were not tied to specific enterprises and their benefits were available to all enterprises regardless of political or administrative connections. Investing in a type of technology or an industrial development base had the same effect. For example, during the Seventh (1986–1991) and Eighth (1991–1996) Five-Year plans, Shanghai invested RMB 120 million in creating a microelectronics research base, and beginning in 1994 the local government began promoting the "Gold Card" Project for computers and software applications.[58] In October 1998, the city began construction of a new $18 million software park in Pudong. The park, a joint venture between the Ministry of Information Industry and the municipal government is intended to attract information technology enterprises to Shanghai. The park will provide infrastructure support to smaller enterprises and has been specifically designed to host data processing, electronic publishing and software assessment, development, packaging, and training centers.

There were, however, distinct patterns to how local research centers were organized in Shanghai, and these patterns helped reinforce the role of the larger state-owned enterprises. Research and development activities in Beijing were concentrated in universities and research institutions; in Shanghai they were concentrated in large- and medium-sized state-owned enterprises. In 1994, Beijing invested over RMB 10 billion in scientific research and development organizations; Shanghai invested 4.29 billion. But during the same period, Shanghai invested six times as much in research and development activities in large and medium-sized, state-owned enterprises than Beijing did. Moreover, Shanghai's local government had a

---

57. "Detailed Analysis of China's IC Production," *China Electronics Daily*, in FBIS-CHI, March 1996, 96–109.

58. "Shanghai Gaoxin Jishu Chanyehua de Chengyin he Wenti" [Success and difficulties in Shanghai's new and high-technology industrialization], *Kexuexue yu Kexue Jishu Guanli* 10 (1993): 21–23.

much larger presence in research institutions than the local government did in Beijing. Of the 251 research institutions in Shanghai, 172 (69 percent) were part of the local government; in Beijing only 20 percent were part of the local government. The remainder had institutional links with the central government.[59]

Focusing R&D funding within the labs of SOEs made the diversion of those funds to worker bonuses or other less productive outcomes more likely. Half of the respondents in a survey of 122 technology enterprises in Shanghai believed that the failure to create special management policies for S&T funds led to their dispersal to non-R&D-related activities.[60] Moreover, Beijing had incentives to foster links between research institutes and production units; it was not providing the operating budget for most research units in the city. The city benefited in terms of taxes and increased employment from successful enterprises. While the same benefits would accrue to Shanghai, the risks were higher. In cases where commercializing new products failed, the city would have to support both the cooperative production enterprise as well as the original research unit.

Direct Funding of Specific Enterprises

Nationally, local governments accounted for between 5 and 9 percent of investment in high-tech enterprises in 1992, and 7 percent of enterprises in the Beijing Experimental Zone listed the city as their supervisory agency.[61] Similar data are not available for Shanghai, and evidence for this type of investment is mainly anecdotal. For example one software company interviewed secured funding from the Jiading branch of the Ministry of Post and Telecommunications (MPT) to develop computer based switching systems. Leaving aside the question of whether or not the enterprise was the best company for the contract, it did not hurt the company's chances of securing the project that its founder had spent three years in the Jiading Science Commission before "jumping into the sea." After the company was founded, its seven-member board was made up of two people from the science commission, two from the local MPT, and three from the original software company.[62]

Given the local government's concern with enterprises in the public sector, local officials were more likely to fund state-owned *minying* enterprises than their more nongovernmental competitors. Local funds were channeled to nongovernmental enterprises that were state-owned, or to joint

59. *Shanghai Kexue Jishu Gongzuo Nianbao: 1995* [Shanghai yearbook of science and technology: 1995], 39–41.
60. Yuan Jianmin, "Shanghai Chuangban Gao Xin Jishu Fengxian Touzi Gongsi de Yanjiu yu Fenxi" [Research and analysis of establishing Shanghai new and high-technology venture capital company], *Shanghai Tongji* [Shanghai statistics] 3 (1993): 10–13.
61. Shilin Gu, *Spin-off Enterprises in China: Channeling the Components of R&D Institutions into Innovative Business*, UNU/INTECH Working Paper No. 16, December 1994, 14.
62. Interview, S11, 16 April 1996.

ventures.[63] In most of these cases, the city government had a presence on the board of directors. This is not to say that there were no small-scale, technologically advanced *minying* enterprises in Shanghai. Rather, alliances between branches of the local government and collective or individual enterprises were concentrated in places like Jiading outside of Shanghai.[64]

The municipal government has more recently attempted to address capital scarcity for *minying* enterprises in a more institutionalized manner. In 1995 the Science and Technology Commission established a development company to invest in critical technologies. The Shanghai Innovative Technology Corporation (Shanghai Chuangxin Jishu Jituan Gongsi) was established with an investment of RMB 31.4 million from the city government and was expected to attract another RMB 9.1 million from foreign investors. Further funds were to be provided by taxes; after 1995, the city government declared that of all taxes collected in the six high-tech zones, 50 percent should be returned to the zones, 50 percent to the R&D development fund administered by the corporation.[65] The Shanghai High and Innovative Technology Development Company, which had branches in all of the city's technology zones and was subsidized by a loan of RMB 1 million from the Economic Commission, funded enterprises with connections to Shanghai Science and Technology University.[66] Two years before Beijing established a similar fund, Shanghai announced in November 1999 the establishment of a RMB 1 billion venture capital fund aimed at boosting the high-technology sector and eventually creating a local Silicon Valley. Much of the investment will be managed by the Shanghai Venture Capital Corporation, which was established by the city government.

Officials in the city's high-technology zones and at the district level also institutionalized high-tech financing. The management office of Caohejing Science Park invested RMB 15 million in the Shanghai Newly Emerging Technology Company (Shanghai Xinxing Jishu Gongsi), of which 5 million was set aside for an experimental venture capital fund.[67] Zhangjiang High-Technology Development Zone sought to solve the problem of venture capital by creating the First Pudong Investment Limited Company

63. The city received less torch funds than Beijing, but data on how the funds were distributed are not available.

64. Interviews, S6 and S7, 6 April 1996.

65. Interview, S37, 23 June 1998.

66. Ding Jun, "Tansuo Gaoke Jiqi Chanyehua de Fazhan Jizhi" [Exploring the operation and development of high-technology industrialization], *Zhongguo Minying Keji Yu Jingji* 4 (1996): 36.

67. "Zuo Yinjin he Chuangxin Xiangjiehe de Fazhan Daolu, Ba Kaifaqu Jiancheng Peiyu he Fazhan Gaoxin jishu Chanye de Jidi" [Take the development road of combining imported technology and innovation, develop and nurture the development zone as new technology industrial base], in *Guojia Gao Xin Jishu Chanye Kaifaqu Suozai Shi Shi Zhang Zuo Tanhui: Wenjian Zailiao Huibian* [Collected documents from the meeting of mayors from cities with national level new and high-technology industry development zones] (Beijing: Guojia Kewei Huaju Jihua Bangongshi, 1996), 95.

(Pudong Diyi Chuangye Touzi Youxian Gongsi).[68] At the district level, in 1995 the district finance bureau and a branch of Jiaotong Bank established the Changpu District Nongovernmental Loan Fund and offered loans of RMB 10 million to five enterprises.[69] The same year Putuo district created an S&T Development Fund of RMB 3.3 million.[70]

Shanghai Enterprises and the Search for Capital

In the face of this capital scarcity, nongovernmental enterprises responded in ways similar to Beijing *minying* enterprises. Benefiting from being the first to enter the market with new products, enterprises relied on income generated from sales. Forward dominated the market with its uninterrupted power supply (UPS) and quickly moved to developing software and application specific integrated circuits (ASIC). But in the information industries, many of the Shanghai nongovernmental companies entered the sector after Beijing companies had already established a market presence. The tax breaks and tax holidays that sustained many of the first nongovernmental enterprises were eliminated in 1993 and enterprises faced an ever-growing array of competitors. Some SOEs became more flexible and entered the computer market, and *minying* enterprises emerged in new regions of development, including Suzhou and Hangzhou.

Moreover, Beijing enterprises benefited from being located in the largest technology market in the country. Companies like Legend had extremely close ties to large potential customers like the Education Commission or the Chinese Academy of Sciences. Legend also benefited from the "computer fever" that gripped many individual consumers in Beijing in the late 1980s. Shanghai *minying* enterprises did not have the same close ties to potential customers; they were cut off from the state-owned enterprises. In fact, many of the enterprises listed as the top Shanghai nongovernmental technology enterprises were branch offices of Beijing enterprises, including Stone, Legend, and Founders.

Many businesses focused on the sale of foreign products. The Zhong Shi Advanced Technology Company concentrated on developing its own financial and management software, but was also Shanghai's first IBM representative. Again, Shanghai enterprises were late to develop these ties and so the market was often dominated by Beijing enterprises. Moreover, the local government directed multinationals into joint ventures with larger, state-owned enterprises. This may have meant an improvement in technology

68. Mai Qiangling, "Gao Xin Jishu Qiye de Fuhuaqi" [New and high-technology enterprise incubator], *Zhongguo Minying Keji Yu Jingji* 4 (1996): 20–21.

69. "Changpu Qu Kewei Wei Minying Keji Qiye Chuangye Danghao: Hao Houqin" [Changpu district science commission does good for nongovernmental enterprise development: Good logistic support], *Zhongguo Minying Keji Yu Jingji* 4 (1996): 43.

70. "Putuo Qu Kewei Wei Minke Qiye Fazhan" [Putuo district science commission in support of nongovernmental enterprise development], 42.

and management skills for SOEs, but it also meant that nongovernmental enterprises did not have the opportunity to learn how to organize technology enterprises or create distribution networks like smaller enterprises with representative agreements in Beijing did.

Enterprises, including Forward and Fuxing, have also diversified into new markets in search of quick gains; Forward moved into health foods, electronic automation equipment, and real estate. Forward also started construction on the "Fuhua High-Technology Park" within Shanghai, the first nongovernmental technology development zone in China. It hoped to collect rental and management fees from domestic and electronics companies that relocate to the site.[71]

PROPERTY RIGHTS: REGULATING AND CLARIFYING OWNERSHIP

Shanghai moved relatively late to regulate and supervise *minying* enterprises. The first Shanghai *minying* enterprises differed little in their property rights structures from those found in other parts of the country. Spun off from local research institutes, universities, or even SOEs, nongovernmental enterprises had unclear boundaries between the ventures themselves and the sponsoring institution. Falling outside of the standard definitions of ownership meant that nongovernmental enterprises in Shanghai faced many of the same ideological attacks levied on enterprises in Beijing. In particular, local cadres and managers of S&T units attacked *minying* institutions in 1984 and 1985 as a misuse of scarce resources, especially skilled personnel. In 1986, Vice Mayor Liu Zhenyuan stated that according to CCP policies, moonlighting workers were not violating any regulations if their work was not affecting their original units. Vice Mayor Liu continued that the original units should allow the workers to go, but that the local government's responsibility was to "stress guidance and management, and a system of moderation that reacts to the fear that moonlighting dissipates the original unit's strength."[72]

In step with this rather lukewarm defense, the city generally moved slowly to regulate *minying* enterprises. In 1985 the Shanghai party committee issued six reports on high-technology development in the city. While not mentioning *minying* enterprises by name, the reports stressed the extreme importance of developing new industries; they also reiterated that new technologies were to be developed in order to reform the ailing state sector. In 1987 the city issued its "Interim Regulations on Developing New Technologies and Industries," which again focused on enterprises that

71. That is, Fuhua was the first zone organized by *minying* principles, not the first zone opened for *minying* enterprises. Interview, S22, 3 September 1996.
72. "Zhe Yang Kan Dai: Keji Renyuan Yeyu Jianzhi" [How to treat S&T personnel spare time and moonlighting], *Keji Ribao*, 28 March 1986.

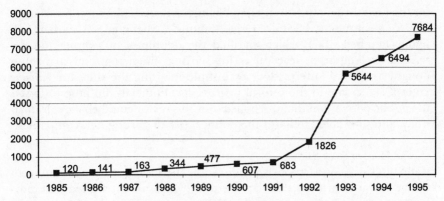

Figure 4.1. Number of Shanghai *minying* enterprises, 1985–1995. *Source: Shanghaishi Keji Tongji Nianjian, 1995* [Shanghai S&T yearbook, 1995].

were under the local government's administrative control. By 1988, Beijing had over 1,800 nongovernmental enterprises with a total income of RMB 1.5 billion, but Shanghai had only 344 enterprises. According to one newspaper report, Shanghai lagged behind because it lacked policies and regulations for *minying* development and local officials failed to provide "systematic guidance" to the sector. Policies in support of S&T development made up less than 10 percent of all new regulations issued during the reform period until 1988. In order to maneuver through the maze of regulations—to receive business licenses as well as establish residence permits for employees—nongovernmental enterprises had to be recognized by local bureaucrats.

The failure to define *minying* meant that enterprises were taxed at the higher rate levied on individual (*getihu*) enterprises. For example, a nongovernmental technology enterprise set up with the support of the local Science and Technology Commission had to pay annual taxes of RMB 63,074, but if the enterprise had registered as an SOE its taxes would have only equaled RMB 16,000.[73] As a result, *minying* entrepreneurs were setting up companies elsewhere; all of the enterprises established by scientists from the Shanghai branch of the Chinese Academy of Sciences were located outside of the city.[74]

It was not until 1988 that the city moved to regulate *minying* enterprises. In May of that year the Science and Technology Commission and the tax bureau created policies that exempted *minying* enterprises from income taxes for their first year of operation and provided a 50-percent reduction

73. Wang Pusheng, "Minban Keji Fazhan Zhong de Wenti Yu Jianyi"[Problems in developing nongovernmental S&T], *Minban Keji* [Nongovernmental S&T] 6 (1991): 8.

74. "Xia yi bu Wang Nar Mai? Guanyu Shanghaishi Shenhua Keji Tizhi Gaige Jige Wenti de Sikao" [Where to next? Some thoughts on deepening the reform of the Shanghai S&T system], *Keji Ribao,* 31 December 1988.

for the next three.[75] In March 1989, Shanghai issued "Measures for Managing Nongovernmental Science Organizations" ("Shanghai Shi Minban Keji Jigou Guanli Banfa"), which clarified the definition of *minying* and described the required procedures for starting, financing, and managing nongovernmental enterprises.[76] By institutionalizing the rules for *minying* enterprises so late in the reform process, local officials did little to lessen the public perception that the policies for this sector could change dramatically, and thus failed to lower the political risk of starting new companies.

The Move to Stock Systems

By the early 1990s, reports produced by the local government and economic research institutes affiliated with the government argued that moving to stock market systems would clarify ownership structures, raise management efficiency, and increase scale of production. In 1992, statistics provided by the Shanghai Science and Technology Commission only acknowledged four types of ownership for nongovernmental enterprises: collective, individual, private, and other. Over 70 percent of all *minying* enterprises were registered as collective.[77] As in Beijing, "collective" was used as an umbrella term for a percentage of private entrepreneurs who registered as collective for political reasons. But in Shanghai between one third and one half of all enterprises registered as collective were not private; in fact they were state-owned enterprises.[78] Not a few SOEs set up nongovernmental enterprises to take care of excess employees and increase workers benefits.

In 1992 the local government began experimenting with forms of stock ownership for technology enterprises. The major goal of these efforts was to convert enterprises of all ownership types into either limited stock companies (*gufen youxian gongsi*) or cooperative stock companies (*gufen hezuo gongsi*). Large, mainly state-owned enterprises were expected to become limited stock companies, enterprises in which all shares were converted into stocks and eventually offered publicly.[79] Huadong Computer, for example, was part of the Bureau of Electronics and was known as the Shanghai Computer Service Company. The 1992 reorganization changed the name of the company, established eight joint subsidiaries (five of which

75. *Zhongguo Minban Keji Shiyejia Xiehui Huixun* [Newsletter of the China nongovernmental science and technology entrepreneurs association], 3 June 1988.

76. *Zhongguo Minban Keji Shiyejia Xiehui Huixun*, 5 May 1989.

77. *Shanghaishi Keji Tongji Nianjian, 1992* [Shanghai S&T yearbook, 1992] (Shanghai: Shanghai Tongjiju Chubanshe, 1993), 208.

78. Mo Gaoku, "Lishun Chanquan Guanxi Cujin Minying Keji Qiye de Jiankang Fazhan" [Clarify property rights relations in order to promote the healthy development of nongovernmental enterprises], *Zhongguo Minying Keji yu Jingji* 5 (1997): 13–14.

79. Xie Lingli, Chen Zhaozhi, Huang Yuemin, and Li Yongjin, *Shanghai Fazhan Yanjiu: Xiandai Qiye Zhidu Lun* [Shanghai development and research: Modern enterprise system theory] (Shanghai: Shanghai Yundong: Chubanshe, 1994), 11.

were joint ventures), and authorized the sale of 40 percent of stocks on the Shenzhen stock market. The board of directors, which was made up of bureau and local Shanghai government officials, owned the remaining 60 percent.

In 1992, the local government allowed the science commission to approve Forward's application to become a joint stock limited company (*gufen youxian gongsi*). The decision was noteworthy since it was the first time a university-based enterprise had been allowed to list on a stock market in China.[80] At the end of the offering, the Forward Group had issued RMB 43 million worth of stock, of which 45,894 shares (46.1 percent) circulated publicly. The two largest shareholders were Fudan University (with 38,665 shares, or 39 percent) and the State Assets Administration Office (with 6,691 shares, or 6.7 percent).[81]

In order to facilitate the clarification of property rights, the Science and Technology Commission created the Shanghai Kehua Invisible Asset Appraisal Office (Shanghai Kehua Wuxing Zichan Pinggu Shiwusuo) in 1994, which was supposed to assess the value of the technology, capital, and services present in technology enterprises. One of the office's missions was to define, evaluate, and thus protect personal investments, but it seems mainly to have operated in defense of the Chinese side in technology joint ventures. Newspaper accounts focus on how foreign partners underestimated the value of joint ventures or overestimated profits and how the office righted those mistakes.[82]

The local government also tried to convert state and collective enterprises to more private forms of ownership. By 1994, the S&T yearbook had expanded the four categories of ownership to ten, but state, collective, and cooperative stock companies, all forms of public ownership, still made up 80 percent of *minying* enterprises. A total of 1,046, or 16 percent of enterprises, had been converted into cooperative stock companies, companies that created internal stock systems in the hope of improving internal management and increasing worker incentives.[83] These internal systems were a transitional step to an eventual public offering, but ownership still rested with a state actor. Those listing stock systems, which offered shares to outside buyers and separated ownership from control, only made up 0.2 percent. In the same period, Beijing made more progress toward clarifying property rights; in 1992 less than 1 percent of all *minying* enterprises were

80. Shanghai Science and Techology Commission, "'Tiao Kuai Mian' Yi qi Gua, Tuijin Shanghai Gao Xin Jishu Chanyehau Fazhan [Grasp the "horizontal vertical" together, promote Shanghai's high-technology industrial development], in *Quanaguo Kexue Jishu Dahui: Wenxian Huibian (xia)* [Collected documents from national science and technology meeting, volume 2] (Beijing: Kexue Jishu Wenxian Chubanshe, 1996), 66.

81. Shanghai Forward Group Limited, *1996 Annual Report*, 8.

82. *Shanghai Keji Ribao*, 19 June 1996.

83. *Shanghaishi Keji Tongji Nianjian, 1994* [Shanghai S&T yearbook, 1994] (Shanghai: Shanghai Tongjiju Chubanshe, 1995), 213.

using some type of stock system. By 1994 the number had risen to 10 percent.[84]

Management Authority within Enterprises

The scale of most *minying* enterprises in Shanghai up until 1993 remained small enough that ambiguity in ownership structures and authority over profits and technology had little impact on how the enterprise was managed. The smallest enterprises were made up of the founder or founders and a few friends, and decisions about technology strategy or hiring and firing remained in the hands of a few. The larger, more successful enterprises created a general manager and a board of directors. Everyday decisions, like hiring and firing, were the general manager's responsibility. Staffed both by members of the supervisory unit and the enterprise, the board was to restrict itself to large investment and general strategy decisions.

Growth created new personnel problems; companies could no longer rely solely on their supervisory agency for their staffing needs. When it started in 1984, all Forward employees came from Fudan University. Until 1992, all skilled personnel and scientists at Forward received salaries from the company, housing and other benefits from Fudan.[85] By 1995, of 350 employees, 90 percent were from outside the university.[86] New employees were mainly on salary and did not receive apartments or the other benefits from Fudan that some older employees did. In order to balance the needs of these different generations, as well as to better react to increased market pressures, Forward began professionalizing its internal management systems. It also adopted an internal shareholding system in order to more closely tie salaries to profits and to create an incentive system that would keep highly trained scientists from leaving the enterprise. Forward distributed 63,200 shares to nine individuals, including the chairman of the board, general manager, director, and several vice general managers.[87]

One of the primary goals of creating stock systems in Shanghai was to clarify questions of ownership and control. In a 1997 report on Shanghai *minying* enterprises, one analyst argued that most enterprises could not actually be called nongovernmental; "management power" (*jingying quan*) remained in the hands of and was controlled by the investing unit. This type of relationship, according to the report, was fundamentally unable to reflect the character of "nongovernmental."[88] Forward used its stock listing as a way to improve and clarify its relations with Fudan. According to General

---

84. Beijingshi Kexue Jishu Weiyuan Hui, *1997 Niandu Beijing Keji Qiye Gongzuo Yaoloan* [1997 overview of Beijing technology enterprises] (Beijing: Zhongguo Jingji Chubanshe, 1998).
85. Interview, S35, 17 June 1998.
86. Chen Suyang, "Forward: Wo de Zhuqiu" [Forward: My pursuit], *Zhongguo Minying Keji yu Jingji* 3 (1996): 38–41.
87. Shanghai Forward Group Limited, *1996 Annual Report*, 8–9.
88. Mo Gaoku, "Lishun Chanquan" [Clarify property rights relations], 13–14.

Manager Chen Suyang, listing on the market allowed the two partners to move from an administrative relationship (*xingzheng guanxi*) to one of capital management (*zichan jingying guanli*).[89] Although Fudan made no initial investment in the enterprise, Forward paid an annual fee of RMB 1 million to the university every year from 1986 until 1992; this payment was not part of a legally binding contract, but was fulfilled every year for the use of Fudan personnel and office space. In July 1992 the annual payment ended, and Fudan was given about 37 percent of Forward's stock, the controlling share. The board of Forward was made up of five individuals, three from the university (the chancellor, vice chancellor, and Communist party representative), two from the company (the CEO and CFO). One manager characterized the board of directors as fairly "hands off" about management, concerned about two main things: the internal distribution of bonuses and management oversight of the CEO.

The new agreement increased the enterprise's autonomy and provided shares worth over RMB 36 million to the university. Still, there were conflicts inherent in the distribution of shares; when the board gave stocks to individuals within the enterprise, part of Fudan opposed, arguing that all of the stocks belonged to the whole university and so individuals should not receive what was in essence collective property. Clarifying these relations, despite the tensions within Fudan, stabilized internal management and made it more likely that the enterprise could hold on to its most talented scientists.

By 2000 most companies had yet to come to an official or formalized agreement over their relations with their supervisory agency. Instead, the agreement was one of administered supervision, in which the enterprise and supervisory unit bargain annually over the share of profits, and the investing unit maintains a degree of control over strategy and internal management. The share of profits varied annually and depends on revenue. This definition of nongovernmental differs fundamentally from that found in Beijing where the agreement is one of negotiated supervision: enterprise and supervisory unit bargain annually over the share of profits, but the enterprise maintained autonomy over strategy and internal management.

GOVERNMENT SUPERVISION:
POLITICAL AND MARKET ACTIVITIES

Local government actors were essential in assisting *minying* enterprises overcome market failures and bureaucratic barriers to continued growth. Key to the sector's success was not only which policies were enacted, but also how they were implemented. As one Shanghai municipal official

89. Next two paragraphs draw from interviews S35, 17 June 1998, and S22, 3 September 1996.

stated, "National policy is all the same, and local leaders will repeat what is being said at the national level about supporting nongovernmental growth. But the most honest of leaders will admit that they have not always done so."[90]

In addition to helping *minying* enterprises surmount problems like underdeveloped markets and unclear regulations, government supervision was essential to encouraging skilled scientists to enter the market in the first place. The Shanghai local government had problems making technology entrepreneurship seem an attractive future. In a 1988 survey of four thousand middle-aged S&T workers, only 12 percent stated they would like to work in *minying* enterprises.[91] The local government's difficulty in creating an environment supportive of entrepreneurship becomes extremely clear in looking at the nongovernmental labor market in Shanghai and Beijing. As mentioned in chapter 3, Beijing was much more successful in encouraging young and middle-aged scientists to start their own companies. In 1996, one third of all *minying* personnel in Shanghai were retired cadres. This has not improved with internet start-ups. In the words of one foreign technology consultant, "Most people would struggle to name more than three or four notable internet entrepreneurs in Shanghai without inadvertently naming a civil servant or government-appointed manager."[92]

FORMAL SUPERVISION: THE SHANGHAI HIGH-TECHNOLOGY
DEVELOPMENT ZONES

City officials institutionalized most formal supervision of technology enterprises through the creation of six high-technology development zones.[93] From the beginning, the Beijing and Shanghai municipal governments approached the development of technology zones differently. Most dramatically, the chronologies differed. The Beijing Experimental Zone was *minying* enterprise led; first nongovernmental enterprises emerged, then the city built the zone. By contrast, the Shanghai local government first created the zones, next it helped create the joint venture firms that would be the vanguard of technological development, and then it helped move other enterprises into the zone. Before construction began on Caohejing, the first zone in Shanghai, the area consisted mainly of farmland. Research organizations like the Shanghai branch of the CAS established offices after the zone opened.

The two cities also had different conceptions of how to organize the

90. Interview, S38, 24 June 1998.
91. "Shanghai Minban Keji Jigou Shige Ri Meng Zeng Bai Yu Jia" [Shanghai nongovernmental institutions rapidly increase to over a hundred], *Wen Hui Bao* [Wen Hui daily], 10 October 1988, 1.
92. Duncan Clark, "Shanghai Lags Internet's Hectic Pace," *China Web*, 11 September 2000.
93. "Zuo Yinjin he Chuangxin Xiangjiehe de Fazhan Daolu" [Take the development road], 95.

zones and how they should relate to *minying* enterprises and to the local economy as a whole. The main mission of Caohejing was to "import newly emerged technology from home and abroad" and "to transfer the achievements of newly emerged technologies to industrial products." While officially the zone welcomed all types of enterprises, multinationals dominated. In 1992, of twenty-seven nationally recognized high-tech zones around the country, Caohejing had attracted the most "three capital" and multinational firms.[94] Nongovernmental technology enterprises were just one type of enterprise in the zones that would help diffuse new technologies to other sectors of the economy.

In addition, the local government differed on the relationship of the zones to the local economy. In Beijing there was much rhetoric that all of the zones in the city were equally important, but the BEZ clearly dominated the local economy. In 1991, the BEZ accounted for 90 percent of all *minying* income, and by 1995 it had only fallen to 78 percent.[95] Shanghai much more clearly encapsulated the idea that the entire city should be the focus of technology policy, not just concentrated areas like science parks. As one science commission official stated: "From the beginning, local leaders wanted to turn the entire city into a zone, not concentrate on one area like Zhongguancun. The whole city has a technological base, the whole city is an economic area, and so the whole city should be the zone itself, not separated from the rest of the economy."[96] As a result, city government control over the zones was extremely decentralized.[97] All six zones in the city were part of the Shanghai High-Technology Development Area. Unlike zones in Beijing or Xi'an, each one of the zones in Shanghai had a degree of independence from the city government. The municipal government created a temporary small working group on the high-tech zones, whose members included Mayor Xu Kuangdi and Hua Jieming, which handled larger questions of direction and policy for the zones. But the group met infrequently, and its last recorded meeting was in 1993.

Below the small group was the Office of the Management Commission of Shanghai High-Technology Industrial Development Zones. The Management Office held nominal supervisory power over the six zones, but each zone had administrative ties with other divisions of the local government. Caohejing, for example, was linked to the Economic Commission, Zhangjiang to the Pudong district government, and Shanghai University Science Park to the Education Commission. The Office of Management acted less as a management office, more as a public relations clearing of-

94. Zhang Zhang, "Huaxia Dadi 'Guigu' Re" [China's "Silicon Valley" fever], *Zhengce yu Xinxi* [Policy and information] 5 (1992): 10–12.

95. Beijingshi Kexue Jishu Weiyuan Hui, *1997 Niandu Beijing Keji Qiye Gongzuo Yaoloan* [1997 overview of Beijing technology enterprises] (Beijing: Zhongguo Jingji Chubanshe, 1998), 81.

96. Interview, S37, 23 June 1998.

97. Following two paragraphs draw from interview, S37, 23 June 1998.

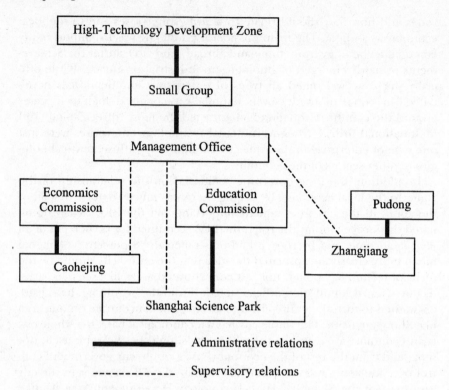

Figure 4.2. Shanghai development zones

fice. One official described their responsibility as follows: "The central government calls us and lets us know that somebody important is coming on an inspection visit, and then we call up Caohejing and tell them they are going to have visitors."

Shanghai's system of zone management had strengths and weaknesses. The main strength of the decentralized system was that it lessened the burden on the city government. Each zone was responsible for start-up costs, including clearing the land for the zone, resettling the farmers who once lived there, and developing all of the new infrastructure for the zone. Each zone had to secure its own loans to develop these projects and was responsible for its own losses. On the negative side, Shanghai's system of management made it almost impossible for the city government to coordinate what all the zones were doing. Attempts to interfere with the internal management of the zones by the science commission were met with resistance by officials who had ties to different branches of the local government. Moreover, each of the zones competed with the other trying to attract enterprises by offering the most preferential policies and thus undermining the local government's attempts to foster certain types of technological development in different areas of the city.

Administration within the Zones: Caohejing

Control and regulation within the zones reflected many of the same strengths and weaknesses of the larger system and can be illustrated with Caohejing. Caohejing's objectives were to meet the component needs of the computer, communication, and consumer electronics industries; by the year 2002, microelectronics, computers, and advance communications were to make up 40 percent of production in the zone.[98] The zone, at its inception in 1984, was considered a local project by the central government, a national project by the local government.[99] A RMB 100 million initial investment, 80 percent of which was put up by the local authorities, established the zone in an area of the city called Caohejing.

Changes within the central government, most notably a reorganization of the State Council, led to the zone's reclassification as a national project in 1988. The reclassification of Caohejing spurred increased financial support from the center and by raising the status of the project, also increased access to more advanced technologies that might have been restricted by foreign exchange limitations. Central funding, even of a project located far from the center, may have actually reduced the ability of the local government to experiment with different ways to organize the zone. As Denis Simon and Detlef Rehn argue, "Due to the considerable financial commitment of both the central government and the Shanghai municipality, one may also expect a stricter level of monitoring and control at each stage of implementation."[100]

A key feature of Caohejing was the role of foreign firms. Preferential tax policies, low land prices, and promises of local government support were used to attract multinationals willing to set up joint ventures and share advanced technologies. Foreign enterprises in Shanghai proper were exempted from local income, land use, and housing taxes for the first three years after their establishment. They were also exempted from paying the local government housing subsidies for Chinese workers.[101] Within the zone, if a joint venture could meet a number of criteria about levels of R&D and percentages of technical personnel on the work force, the enterprise was eligible for more preferential treatment.[102] By 2000, 210 foreign firms had invested a total of $1.2 billion in the zone.[103]

98. Shangahi Gao Xin Jishu Kaifaqu Guanweihui Bangongshi [Management office of the Shanghai new and high-technology development zone], *Shanghai Gao Xin Jishu Chanye Kaifaqu "9.5" Guihua ji 2010 nian Zhanwang* [Forecast for Shanghai high- and new technology development zone for Ninth Five-Year Plan and 2010] (Shanghai, 1998), 17.

99. Denis Fred Simon and Detlef Rehn, *Technological Innovation in China: The Case of the Shanghai Semiconductor Industry* (Cambridge, Mass.: Ballinger, 1988), 137.

100. Simon and Rehn, *Technological Innovation in China,* 138.

101. Shanghai Municipal People's Government, "Provisions of the Shanghai Municipality for the Encouragement of Foreign Investment," 23 October 1986.

102. Shanghai Municipal People's Congress, "Interim Regulations on Shanghai Caohejing Hi-Tech Park."

103. Shanghai Caohejing Hi-Tech Park, available online at http://www.shtp.com.cn/1/jjgk. htm, accessed May 2001.

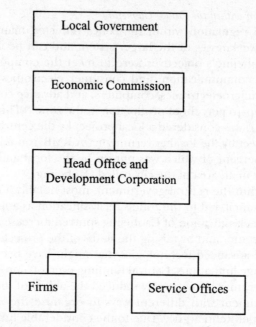

Figure 4.3. Management of Caohejing technology park

Within Caohejing, the zone's management organization was equal in administrative standing to the profit-seeking branch of the zone.[104] Officials described the system as "one company, two different signs on the door" (*yi ge gongsi, liang ge paizi*).[105] The management office administered the zone for the city, but was equally responsible for real estate, construction, and investment opportunities that the office pursued for profit.[106] Advantages of this type of organization included the injection of profit motive and market competition into the running of the zone itself. Yet the general office lacked the administrative powers necessary to coordinate relations between various actors and to break down barriers between bureaucracies.

Supervising Nongovernmental Growth in the Zones
Beginning in 1994, there was a growing realization that Shanghai's development zones were not serving the needs of *minying* enterprises, and that Shanghai needed to promote a system of innovation more reliant on nongovernmental enterprises. In 1995 the State Council stopped differentiating between nationally and locally approved zones, ending the prefer-

104. Compare to Beijing, where the Strong Corporation is below the management office.
105. Interview, S2, 17 March 1996.
106. Li Ping, "Guo Nei Wai Gaojishu Kaifayuan Guanli Moshi de Bijiao Yanjiu" [A comparative study of the management systems of foreign and domestic high-tech development parks], *Kexuexue Yanjiu* [Research in the sociology of science] 12, 4 (1994): 74.

ential policies that national zones like Caohejing received and the nationally defined criteria they had to meet to be eligible for those benefits.[107] The dismantling of this system allowed greater local variation in how science parks were managed.

The Shanghai Science and Technology Commission designated Jiading as a special area in which to promote *minying* development and projected that by 2002 fifty high-technology enterprises would have total industrial output value over RMB 5 billion.[108] Within the city's high-tech zones, the city helped create "innovation centers" (*chuangye zhongxin*) and "incubators" (*fuhuaqi*) for *minying* enterprises. The role of incubators included business education, managing personnel dossiers, and product development assistance.[109] In addition, a special consulting company that deals with bureaucratic pitfalls was established to assist scientists and students returned from study abroad who want to set up their own high-tech enterprises; they were also offered low rents and discounted real estate prices.[110]

Informal Supervision

In Shanghai, informal supervision was much less extensive and much less focused on *minying* enterprises than in Beijing. While the local government did sponsor a number of citywide meetings of S&T personnel, there was no equivalent of the specialized meetings held for *minying* entrepreneurs in Haidian in 1984 and beyond. In 1993 the city established a consultative committee for S&T development that included Vice Mayor Xu Kuangdi and twenty-six members of the Academy of Sciences. It was not until 1995 that the city held a meeting about *minying* enterprises.[111]

Even when the government did hold meetings, there was a sense among *minying* entrepreneurs that local officials did not understand the rules of the game for high technologies. As one entrepreneur stated, "Local officials repeat what they are supposed to say about high-tech development, but they do not truly understand."[112] Shanghai entrepreneurs were aware of the dense social and personal connections that seem to characterize high-

107. Interview, S37, 23 June 1998.
108. Shangahi Gao Xin Jishu Kaifaqu Guanweihui Bangongshi, [Management office of the Shanghai new and high-technology development zone], *Shanghai Gao Xin Jishu Chanye Kaifaqu "9.5" Guihua ji 2010 nian Zhanwang*, 14.
Jiading is within Shanghai prefecture, about forty-five minutes north of the city. In 1996 the Baoshan Science and Technology Commission established the Baoshan nongovernmental technology park northeast of the city, but the zone is not included in discussions of Shanghai's technology plans. Interview, S32, 12 June 1998.
109. Mai, "Gao Xin Jishu Qiye de Fuhuaqi" [High-technology enterprise incubator], 20.
110. "Shanghai Chuxian 9 ge Keji Qiye 'Fu Hua Qi'" [9 high-tech enterprise "incubators" emerge in Shanghai], *Wen Hui Bao*, 18 November 1996.
111. "Shanghai Sets Up S&T Consultative Body," *Wen Hui Bao*, 14 July 1993 in FBIS-CHI 145–16–1993, and "Shanghai Minying Keji Jinru Kuai Che Dao" [Shanghai nongovernmental S&T is soon to arrive], *Minban Keji* 8 (1995): 33.
112. Interview, S18, 22 June 1996.

tech development; they used scientists in both Silicon Valley and Beijing as reference groups. Yet it was seen to be particularly difficult to develop social relations in Shanghai, either with local officials or with other entrepreneurs. One interviewee complained that those in local government, especially those in the planning commission, knew nothing about technological development. This was understandable, given their position. What was worse, according to this entrepreneur was "that the people in the science commission did nothing for us either: no meetings, no dinners, and no exchanging ideas."[113]

It is in the realm of informal social and political functions that the Shanghai authorities were particularly deficient. In the early 1980s, local officials did speak out in defense of S&T personnel's right to moonlight and to look for new jobs outside of the state sector, but most entrepreneurs in Shanghai spoke of a continued sense of "social discrimination." Residents did not view *minying* enterprises as a respected career path, and local officials did little to change that view.[114] While Beijing officials worked to improve the social standing of *minying* entrepreneurs, praising their high moral quality and their contributions to the local economy, Shanghai officials equated them with individual (*getihu*) enterprises. The general manager of Fuxing, Guo Chanchang, argued that "we are not just a *getitu* enterprise interested in making money. We are a group of young people with ideals and aspirations. Our goal is to strive for the honor of China's nongovernmental enterprises."[115] These views were not widely held in the municipal government and one author from the Commercial and Industrial Bureau wrote that, like private enterprises, *minying* enterprises often arbitrarily changed their business addresses, embezzled money, and withheld employee salaries.[116]

In a strange twist to what tended to happen in Beijing, a local science commission officer in Changning district discussed her attempts to raise the social status of science entrepreneurs. "It took a while," she stated, "but I finally convinced the other cadres that the quality of *minying* entrepreneurs was quite high."[117] In Beijing, cheated customers denigrated the morals of high-tech entrepreneurs and local officials worked to assure the public that nongovernmental entrepreneurs were good citizens; in Shanghai, science and technology commission officials had first to convince other government officials.

Again, as with the zones, there was a shift, at least at the district level, in

---

113. Interview, S11, 16 April 1996.
114. "Shanghai Minban Keji Jigou Shige Ri Meng Zeng Bai Yu Jia" [Shanghai nongovernmental institutions rapidly increase to over a hundred], *Wen Hui Bao*, 10 October 1988.
115. "Shanghai Minying Keji Qiye de Jueqi he Fazhan" [Rise and development of Shanghai nongovernmental technology enterprises], *Zhongguo Keji Chanye Yuekan* 9 (1995): 29–30.
116. Wang Enshou, "Minban Keji zai Shanghai" [Nongovernmental S&T in Shanghai], *Minban Keji* 6 (1992): 33.
117. Interview, S16, 24 April 1996.

focus toward *minying* enterprises in the 1990s, a move from suspicion to service. Changning district adopted preferential polices earliest, creating a bonus fund for S&T personnel and providing individual tax breaks of up to 20 percent. By 1994, nineteen of the top one hundred *minying* enterprises in Shanghai were based in the district.[118] In Changpu district, local science and technology officials argued that "nongovernmental enterprise development cannot be separated from governmental support," and the district passed preferential tax policies and improved coordination of *minying* enterprises.[119] Local officials in Putuo wrote research reports, made registration easier, and even organized study trips for *minying* entrepreneurs to Zhongguancun and the United States.[120]

At the city level, local officials also began to address the problems of small enterprises. In September 1999, Shanghai released regulations covering company establishment, administration, and operation intended to create a better social environment for small enterprises. The new regulations allowed small companies to pay in installations to register capital and promised a two-week processing time for official registration.[121] In addition the city announced its intentions to help small- and medium-sized enterprises, especially high-technology enterprises, gain access to funding. The city also planned to establish a loan fund to act as a guarantor when small enterprises seek bank loans.[122]

What the long-term impact of these policies will be remains in question, but Shanghai's pattern of supporting large enterprises over small ones seems to be reproducing itself in the internet. Government involvement in Shanghai—through control of infrastructure, access, content, and an extensive licensing regime—is much more pervasive than in Beijing, Shenzhen, or Guangzhou.[123] Shanghai was the first mainland city to use integrated service digital networks (ISDN) and pioneered the installation of high-speed asymmetric digital subscriber lines (ADSL) connecting offices and homes. The flip side of this government support for infrastructure is extensive involvement in access and content as well as a tight regulatory regime. Shanghai Online, operated by Shanghai Telecom, dominates internet access in the city. Offering a free dial-up service for users to access sites hosted on its server, Shanghai Online has also managed to dominate con-

118. Wu Weikang, "Zhaoshang You Shu, Liuke You Fang" [Skilled in attracting and retaining business], *Zhongguo Minying Keji Yu Jingji* 4 (1996): 43. See also "Jingan Qu Minying Keqiye Jinying Guimo You Tigao" [Jing An district improves management of nongovernmental enterprises], *Shanghai Keji Ribao*, 29 March 1996.

119. "Changpu Qu Kewei Wei Minying Keji Qiye Chuangye Danghao: Hao Houqin" [Changpu district science commission does good], 43.

120. "Putuo Qu Kewei Wei Minke Qiye Fazhan" [Putuo district science commission in support], 42.

121. "Shanghai Issues New Policy to Promote Small Enterprise," *China News Digest*, 19 September 1999.

122. Zeng Min and Huo Yongzhe, "More Policy Support for Small and Medium Enterprises," *China Daily*, 14 January 2000.

123. Clark, "Shanghai Lags."

tent, something China Telecom has not been able to achieve elsewhere in China.

CONCLUSION

At the beginning stages of reform, policymakers in Shanghai held a conception of technological development distinct from that found in Beijing. Unlike in Beijing, relatively autonomous enterprises were not to drive the city's technological development; rather technological development centered on state-owned enterprises and large-scale production in general. A whole range of technology policy decisions from how to commercialize new R&D products to how to manage high-technology development zones were tailored to reinforce the position of SOEs in the local economy. Science funding was directed to labs within SOEs; local officials moved slowly to regulate newly founded technology enterprises; and the high-technology zones offered significant preferential policies to large MNCs, but generally ignored smaller start-ups.

Minying enterprises suffered not so much from active resistance, but more from neglect. Local officials did not push *minying* enterprise development when the center encouraged them to do so, and perhaps more important, they did not encourage high-technology entrepreneurs when the center was silent on the issue. Essential to *minying* development were a whole range of informal supervisory roles that local officials did not adopt until much later in the reform process.

These policies had both a direct and indirect impact on enterprise organization and behavior. Compared to Beijing, local government action created a much more hierarchical pattern of technological development. The majority of enterprises involved in high-tech industries were large, state-owned (though often listed as collective or cooperative stock), and had strong ties to the municipal administration. Moreover, financial and regulatory policy decisions affected entrepreneurs' ability to build social and economic networks that may increase the competitive ability of individual enterprises. The pattern of development in Shanghai left most small enterprises isolated from each other and from local government actors, and the local government had little success in coordinating action between actors across different administrative systems.

Lacking in Shanghai during the 1980s and early 1990s was a network of semi-independent research institutes, *minying* associations, and individual entrepreneurs that linked enterprises to the local government. There was little process of interaction or feedback between state and nonstate actors. There were no equivalents of management concepts like the "four self principles" (*si zi*) or the "second breakthrough" (*er ci chuangye*) developing within enterprises and then being popularized and reduplicated by the local government. In fact, Shanghai entrepreneurs and officials used the same terms developed in Beijing to describe their development strategy.[124] The Shanghai Enterprise Association for Science and Technology (Shanghai Shi Keji Qiye Lianhe Hui) is a top-down organization sponsored by the Science and Technology Commission that represents more than 160 enterprises of all shapes and ownership structures.[125] Although the Shanghai Economic Development Research Institute (Shanghai Jingji Fazhan Yanjiusuo) has done a few projects on the *minying* sector, there is no equivalent in Shanghai to Beijing's Great Wall Research Institute, a company that undertakes both management consulting for *minying* enterprises and research projects for the local science commission.[126]

Municipal leaders have begun to pay more attention to smaller enterprises, and Shanghai's role as a financial and commercial center may eventually be a more important factor in determining the future trajectory of nongovernmental enterprise development than the relative lateness of government supervision. Close relations with potential end-users are essential for innovative enterprises and their understanding of market potential. Beijing's role as a cultural and political center and its lack of industrial enterprises could handicap enterprises in the capital as they enter a more mature stage of development. If nongovernmental enterprises in Shanghai can create links to the rapidly growing local economy, they have both the scientific knowledge and organizational resources to enter a new stage of growth.

124. See for example the report produced by the Shanghai Science and Technology Commission that states *minying* enterprises are entering the "second breakthrough." Shanghai Shi Kexue Jishu Weiyuanhui, *Jueqi de Shanghai Keji Qiye* [Rising up of Shanghai science and technology enterprises] (Shanghai, n.d.), 2.

125. Interview, S38, 24 June 1998.

126. Interview, S34, 17 June 1998.

CHAPTER FIVE

Nongovernmental Enterprises in Guangzhou and Xi'an

There may be no two cases of technological development in China farther apart than the cities of Guangzhou and Xi'an. Across almost every measure, the two cases seem to be the reverse image of each other. As the capital of Guangdong, Guangzhou benefited from changing international markets, improved means of communication, and geographical closeness and historical ties to Hong Kong. Moreover, reformist provincial leaders aggressively exploited the "policy windows" opened by the central government on behalf of the region, often pushing reforms much further than had originally been expected and fostering some of the world's fastest growth rates in the 1980s and 1990s.

Xi'an, capital of Shaanxi, had neither geographical advantages nor preferential policies. Large distances and inadequate roads and railways as well as its location in China's interior isolate Shaanxi from the rest of the country as much as from foreign investors. Throughout the 1980s, the center actively promoted a development policy that granted preferential treatment to the coastal areas to "get rich first" at the same time that economic development in the interior regions was deferred. In addition, provincial leaders were slow to take advantage of the policy openings that were granted to them; Shaanxi trailed behind much of the country in agricultural reforms, the introduction of the management contract system in state-owned enterprises, and the support of town and village enterprises.[1] Between 1978 and 1986, Shaanxi's share of combined national output in agriculture and industry declined steadily, dropping the province from seventeenth to twenty-fifth among twenty-nine listed provinces.

Differences at the provincial level reverse themselves at the municipal level. While the general pace of reforms was slow in Shaanxi, Xi'an dominated the province's economic, political, and cultural life; Xi'an has 80 percent of the province's S&T resources and industrial foundation. North and south Shaanxi are sparsely populated, and economic activity occurs in the shadow of Xi'an, its large state-owned enterprises (SOEs), many of

---

1. Xian Dongfang and Cheng Huigang, "Shaanxi Jingji Fazhan Mianliande Kunjing yu Xuanze" [Difficulties and choices that Shaanxi faces in economic development], *Jingji Gaige* [Economic reform] 1 (1988): 39.

which belong to the defense sector. Perhaps most important, at the beginning of the reform period, Xi'an had the country's third largest concentration of science and technology resources.[2] By contrast, Guangzhou, dominated by many older, labor-intensive, small-scale enterprises, did not keep up with cities like Shenzhen and Zhuhai in Guangdong.[3] Economic life slowly migrated south, and one official compared the provincial capital with its outdated industrial base and overcrowded neighborhoods to a "tired old man."[4] Moreover, during the 1980s and 1990s, Guangzhou trailed the rest of the country in the number and quality of research institutes, universities, and trained science and technology personnel. In 1985 Guangdong ranked twenty-third in S&T personnel as percentage of total population, and in 1993 science and technology personnel made up only 15.3 percent of total employees in high-technology enterprises in the city.[5]

Comparing these two cases, even with their differences, highlights the difficulties faced by all but the most developed and policy-adept localities in fostering indigenous nongovernmental high-tech enterprises. Few regions possessed both the factor endowments and political skills necessary to support technological development. One entrepreneur who returned to Xi'an from Guangdong to start his own company characterized the growth of nongovernmental enterprises in the two cities this way: "In Xi'an, *minying* enterprises rely on technological capability and talented individuals. Guangzhou enterprises depend on markets, on trade, and on Hong Kong."[6] Yet having just "technology push" or "market pull" was not enough to ensure technological capability. Rather, local governments needed to put all three pieces of the puzzle together. That is, nongovernmental enterprises required technological capability, access to capital, and policies supportive of *minying* enterprise growth.

Neither Xi'an nor Guangzhou managed to combine all three; Xi'an had technology push, Guangzhou market pull, and both were late to develop policies supportive of high-technology enterprises. Local leaders in

2. Xi'an's third place spot in the S&T infrastructure, behind Shanghai and Beijing, is frequently mentioned in interviews, policy documents, and articles about the city. Liu Yin and Li Huazu provide "mathematical proof" of this with an equation that includes number of research units, population, patents, and research personnel in SOEs. See "Xi'an Diqu Liyong Gao Ke Jishu Xi Yin Waizi de Sikao" [Some thoughts on Xi'an using high technology to attract foreign investment], *Kexuexue yu Kexue Jishu Guanli* [Sociology of science and S&T management] 10 (1993): 27–29, 32.

3. By 1993, 12.7 percent of Shenzhen's industrial output was considered "high-technology," 7 percent higher than the national average. See *Xinhua*, 27 November 1993. Shenzhen also accounted for 40 percent of provincial high-tech products.

4. Quoted in Ezra F. Vogel, *One Step Ahead in China: Guangdong under Reform* (Cambridge, Mass.: Harvard University Press, 1989), 196.

5. "Guangdong Keji Renyuan Liudong Quxiang, Shenghuo Daiyu Piandi Xianxiang Yiran Cunzai" [Trends in Guangdong S&T personnel flows, low salaries still exist], *Keji Ribao* [Science and technology daily], 18 January 1988; and Wu Yuezheng, "Guangzhou Gao Xin Jishu Chanye Pinggu Ji Duice" [Countermeasures and evaluations of Guangzhou new and high-technology industries], *Tansuo* [Exploration] 6 (1993): 22–25.

6. Interview, X3, 23 July 1998.

Guangzhou and Xi'an tended to focus on only one of the two descriptive couplets: either nongovernmental or high-technology. For Guangzhou leaders it was enough to encourage the growth of foreign investment and private enterprises; technological capability, they believed, would eventually be transferred to domestic producers. Incentives were created for investing in joint ventures (JVs) or town and village enterprises (TVEs), but not high-technology enterprises. In 1992, Guangzhou led the country in overall private enterprise growth, and by 1999 private enterprises accounted for 31.7 percent of high-tech enterprises in the province.[7] But *minying* enterprises remain starved of capital and unable to hold on to skilled personnel.

Xi'an, much like Shanghai, could not and did not encourage *minying* enterprises in a world of state-owned enterprises. While Guangzhou led the country in private enterprises, Xi'an ranked tenth.[8] Provincial and municipal leaders hesitated to take the lead in any area of reform.[9] A 1994 *Enlightenment Daily* article looking at the history of reform during the 1980s asked why, with all of its S&T resources, Shaanxi trailed so far behind the rest of the country in economic development? The short answer, according to the article, was limited vision and conservative leadership.[10] Local leaders essentially ignored *minying* enterprises in the first stages of reform, leaving them unrecognized by the economic bureaucracy.

In addition to demonstrating the difficulties faced by all but the most advantaged localities, these two cases also illustrate the process of learning and policy diffusion in technological development. Compared to Shanghai and Beijing, Guangzhou and Xi'an were late developers. *Minying* growth in both locations did not take off until after 1993 and since then has trailed Beijing and Shanghai. By 1995 there were only four thousand nongovernmental enterprises in Xi'an and eleven hundred in Guangzhou compared to more than seven thousand in Shanghai.[11] In Guangzhou, technology enterprises had few ties to each other and tended to get lost in an ocean of private enterprise. In Xi'an, enterprises relied more on developed personal

7. "Review of Enterprise Reform in Provinces," *Nanfang Ribao*, 16 August 1999, in *Foreign Broadcast Information Service-China* (hereafter FBIS-CHI), 20 August 1999.

8. Li Beihai, "Xi'an Siying Jingji Saomiao" [A scan of Xian's private economy], in *Zhongguo Jingji Minyinghua Yanjiu* [Research on privatization of China's economy], ed. Zhang Qikai (Xi'an: Xibei Daxue Chubanshe, 1995), 215.

9. Hence Kevin Lane's characterization of provincial leaders as "one step behind." See his "One Step Behind: Shaanxi in Reform, 1978–1995," in *Provincial Strategies of Economic Reform in Post-Mao China: Leadership, Politics, and Implementation*, ed. Peter T. Y. Cheung, Jae Ho Chung, and Zhimin Lin (Armonk, N.Y.: M. E. Sharpe), 212–52.

10. Ling Xiang, "Shaanxi Keji Lingxian Yuanhe Jingji Zhizhou?" [Shaanxi leads in science and technology, why does it trail behind economically?], *Guangming Ribao* [Enlightenment daily], 6 December 1994.

11. Interviews, X2, 22 July 1998, and G15, 6 July 1998. Also see, "1995 Nian Minying Keji Qiye Fazhan Taishi" [1995 nongovernmental technology enterprise development trends], *Zhongguo Keji Chanye Yuekan* [Chinese technology industry monthly] 8 (1996): 40.

networks and slowly worked themselves toward the center of a local economy dominated by unproductive SOEs.

After 1992 Guangzhou and Xi'an learned from and began to adapt the experiences of Beijing and Shanghai to local conditions. Local officials in both Guangzhou and Xi'an devoted more resources to *minying* enterprise growth. In Guangzhou, local officials expressed growing skepticism that foreign direct investment alone could raise the technological level of local factories. In Xi'an, continued frustration with the slow pace of SOE reform and the success of a few local *minying* enterprises attracted the attention of city officials.

LOCAL DEVELOPMENT STRATEGIES AND TECHNOLOGICAL INNOVATION

In fostering the development of information industries, local leaders had to decide how best to organize and coordinate a wide range of economic and technological resources, actors, and institutions. Local governments had to choose which technologies were critical to the economy and what the government's role should be. Facing a different set of constraints and opportunities, municipal leaders in Guangzhou and Xi'an adopted different development strategies and development paths.

Guangzhou: Asia's Fifth Dragon
Beijing, influenced by the success of Silicon Valley, promoted a definition of *minying* that centered on autonomous and technologically advanced enterprises. Guangzhou's inspiration, however, came from a different and closer source. In particular, Guangzhou looked to the experiences of the "Four Dragons": Korea, Hong Kong, Taiwan, and Singapore. According to Guangzhou planners, the experience of these countries could be broken down into four lessons.[12] First, growth and development cycles have been dramatically shortened; what took one hundred years to accomplish in the West, Asian countries did in fifty. Second, after a short period of import-substituting industrialization, a country should take advantage of growing international trade and shift to export-oriented industries. Third, technology was essential; the host country must raise its own research and development capabilities by using licensing agreements and importing the most advanced technology. Finally, export-led growth and technological imports need to be accompanied by infrastructure investments, especially in education and human capital.

Many of these lessons were behind the policies adopted by the central

12. See for example an article by the Research Office of the Guangzhou Statistics Department, "Guangzhou Ganshang Si Longtou Chu Tan" [Discussion of Guangzhou catching up with the Four Dragons], in *Guangzhoushi Youxiu Tongji Fenxi Xuanbian* [Selected analysis of Guangzhou statistics] (Guangzhou: Guangzhou Tongjibu Chubanshe, n.d.), 2–3.

government from 1986 to 1988 known broadly as the coastal development strategy. These policies decentralized authority over trade, finance, and taxation to local governments, allowed rural enterprises to form joint ventures with foreign partners, and provided preferential policies to foreign investors.[13] In effect, these policies made Guangzhou and other southern cities a more welcome site for labor-intensive manufacturers from Taiwan and Hong Kong who had lost their competitive advantage at home. Local leaders in Guangdong and Guangzhou, eager to begin building much needed basic infrastructure and addressing long dormant social welfare needs, aggressively pushed these policies, encouraging the inflow of foreign capital.

Within this larger export strategy, creating an indigenous technological capability was either assumed to be an organic part of opening to the outside world or subsumed to the desire to attract the most foreign investment possible. Importing advanced technologies was seen as the flip side to developing China's own S&T products.[14] Yet, in reality the coastal development strategy eventually abandoned attempts to restrict investment to projects that promised high levels of technology transfer. Earlier policies, including the creation of Special Economic Zones (SEZs) in Shenzhen and Zhuhai, had nominally restricted foreign participation to projects providing advanced technologies.[15] With the adoption of the coastal development strategy, officials welcomed low-technology, labor-intensive manufacturing based on the promise of foreign exchange.

According to Vice Governor Lu Zhonghe, the drive to develop an export-oriented economy had to be balanced by policies to attract and absorb talented individuals to the province. The "wing" of encouraging S&T personnel to establish their own enterprises balanced the "wing" of TVE development, creating a virtuous cycle of technological development and export-oriented growth.[16] Guangzhou Mayor Zhu Senlin put it even more simply: "Developing an export economy without improving science and technology is impossible."[17] In reality, however, nongovernmental technology enterprises as separate entities were, for the most part, either overlooked or unacknowledged in Guangzhou. Unlike in other parts of the

13. See Barry Naughton, "Economic Policy Reform in the PRC and Taiwan," in *The China Circle: Economics and Technology in PRC, Taiwan, and Hong Kong*, ed. Barry Naughton (Washington, D.C.: Brookings, 1997), 81–110.

14. Or as one newspaper account described the strategy, "using a bought pot to boil your own rice" (*yong mai laide guo zhu ziji de mifan*); "Keji Tizhi Gaige Dayou Gaotou" [Reforms of the S&T system have a big start], *Keji Ribao*, 5 April 1987.

15. George Crane, *The Political Economy of China's Special Economic Zones* (Armonk, N.Y.: M.E. Sharpe, 1990), and Fuhwen Tseng, "The Political Economy of China's Coastal Development Strategy: A Preliminary Analysis," *Asian Survey* 31 (March 1991): 274.

16. "Gongmao Jiehe, Yikao Keji, Gaohuo Jingji" [Combine industry and trade, rely on S&T, enliven the economy], *Keji Ribao*, 29 March 1988.

17. Ibid.

country, local leaders tended to lump *minying* together with private (*siying*), underplaying the industry-specific needs of nongovernmental entrepreneurs. As one Guangzhou official exclaimed, "Everything that is not state owned is private. If nongovernmental means no outside interference, well then we are all *minying* in Guangzhou."[18] On one hand, this statement reflects how far reforms had gone in Guangzhou. Private ownership was not as politically sensitive for technology or any other type of enterprise. In fact, as early as 1993 Guangzhou nominally put private and state enterprises at the same level by eliminating controls over private businesses in regard to scope, credit, and finance.[19]

On the other hand, the rush to foreign capital, TVEs, and private enterprises meant that little attention was paid to the difficulties of creating an enterprise system dedicated to developing high technologies. In contrast to Beijing's efforts to support organizationally advanced enterprises and then to protect them from outside interference, Guangzhou leaders were content to open the policy window as wide as possible but do little for technology enterprises. This meant that a local government more focused on attracting foreign direct investment was slow to create the institutions necessary for absorbing and diffusing technological capabilities to domestic producers.

Even at the time of its adoption, the coastal development strategy generated criticism.[20] At the national level, analysts exhibited skepticism toward an overreliance on this development strategy. At the local level, leaders worried that the coastal development strategy would leave Guangzhou without the tools to develop its own high-technology enterprises. At a meeting of the provincial and city science and technology commissions held in Guangzhou in 1988, delegates complained that reforms of the S&T system were not keeping up with the city's economic needs. Technology imports had already consumed $5 billion, but the equipment imported was often too advanced and only 1 to 2 percent of that equipment was actually being put to productive use in domestic enterprises. Moreover, scientific planning was often sidetracked from the more difficult path of fostering innovation by more short-term, profit-oriented goals.[21]

These fears were further focused for local leaders in Guangzhou by

18. Interview, G5, 12 November 1996. In a book produced by the Guangzhou Academy of Social Sciences and titled *The Charisma of Guangdong's Ten Large Nongovernmental Enterprises,* "nongovernmental" only appears on the title page. Throughout the book, enterprises are called private (*siying*). See Yan Linsen, *Guangdong Shida Minying Qiye Fengcai* (Guangzhou: Huacheng Chubanshe, 1998).

19. *Xinhua,* 23 July 1993.

20. Wei Dakuang and Gao Liang, "Ideas on the Strategic Economic Concept of the 'Great International Cycle,' " trans. in *Chinese Economic Studies* 25, 1 (fall 1991): 16.

21. "Guangdong Zheng Xie Weiyun Hui yu Zhengdun Keji Zhixu" [Guangdong political consultative committee appeals for a reorganization of the S&T system], *Keji Ribao,* 22 October 1988.

changes in both external and internal markets.[22] During the early 1990s, China lost some of its comparative advantage in labor-intensive manufacturing; rising land and labor costs convinced some foreign producers to relocate to Malaysia, Indonesia, or Thailand. In addition, the central government ended the city's comparative advantage in preferential policies; in 1992 interior provinces were allowed to offer many of the same incentives to attract foreign investment. Finally, Guangzhou faced growing competition from Shenzhen, which had adopted a more comprehensive strategy to develop its own innovative technology enterprises.

Technological Innovation in the Interior: Xi'an

In supporting *minying* enterprise development, Xi'an has been pulled between the different poles of technology policy reflected in the Shanghai and Beijing models. Like Shanghai, accounts of technological development from the mid and late 1980s began with descriptions of the decrepit state of Shaanxi's industrial structure; metalwork, aviation, and defense technologies, for instance, were all saddled with equipment from the 1960s and 1970s.[23] Local planners also stressed the dominant role of large state-owned enterprises in the economy and the need to link high-technology development with the revitalization of more traditional industries.

Xi'an municipal officials, like those in Shanghai, also focused on the role their city played in a larger regional economy. Planners in Xi'an were concerned with the needs of state-owned enterprises outside of their city and their province. Technological development was expected to radiate first north and south out of Xi'an to the poorer regions of the province, and eventually to neighboring provinces.[24] The small size of *minying* enterprises meant they could be of little help in attaining the city and province's developmental goals. As one academic involved in technology reform argued, "During the 1980s, local leaders focused all their attention on SOEs, especially those involved in the defense industries. They could barely see, much less help, nongovernmental enterprises."[25] At best, large *minying* enterprises could become partners in large industrial groups. In 1995 the leaders announced the "1851" plan for coordinating economic and technological economy in the province. The plan's goals included fostering a

---

22. Kuang Zhiyun, "Dali Fazhan Guangdong Sheng Gao Xin Jishu Chanye" [Energetically develop Guangdong province high and new technology industries], *Jingji Guanli* [Economic management] (1995): 12–15.

23. "Shaanxi Yinjin Jishu Cheng Liang Xing Xunhuan" [Shaanxi's advanced technology enters a virtuous cycle], *Keji Ribao*, 20 November 1987.

24. Shaanxi Province People's Government, "Fazhan Minying Qiye, Peiyu Xinguang Chanye, Jiasu Shaanxi Jingji Fazhan" [Develop nongovernmental enterprises, cultivate new industries, speed Shaanxi's economic development], in *Quanguo Kexue Jishu Dahui: Wenxian Huibian* [Documents from the national meeting for science and technology] (Beijing: Kexue Jishu Wenxian Chubanshe, 1995), 152.

25. Interview, X7, 24 July 1998.

new high-technology industrial belt in Xi'an, developing fifty high-tech products with national competitiveness, and forming ten large enterprise groups.[26]

As time passed, however, the province veered closer to the Beijing model and its greater focus on smaller enterprises. Local leaders grew increasingly aware that reform of the SOE system was not going to be easy or quick; "Revitalizing large state-owned enterprises is an important part of deepening reform. But it will not be completed in one day."[27] In 1991 provincial vice governor Jiang Xinzhen expressed the need to provide support for *minying* enterprises; "Within the entire S&T system, nongovernmental enterprises are an area we cannot lack."[28] The success of science-based enterprises in Beijing also attracted attention in Xi'an; local leaders began to ask themselves why Xi'an had not taken better advantage of its abundant S&T resources. Or as officials in the management office of the Xi'an high-tech zone put the question somewhat more colorfully, "Why did Xi'an have a golden bowl, but still have to go begging for food?"[29]

In order to put an end to this begging, local leaders began constructing an alternate development strategy. City officials sent study groups abroad and to more established zones (including Beijing) to learn how to better support innovation.[30] In 1992 and 1993 the city and provincial governments issued numerous regulations concerning the definition and regulation of nongovernmental enterprises. By 1995, the Shaanxi provincial government could make a legitimate claim that it was actually pursuing its declared strategy toward *minying* enterprises of "grasping the big, raising the small" (*zhua da, yu xiao*), or developing large enterprises and helping increase the scale of small ones.[31]

26. "Shaanxi Governor Stresses High-Tech Industries," *Shaanxi Ribao*, 21 August 1995, in FBIS-CHI, 21 August 1995.

27. *Zhongguo Gao Xin Jishu Chanye Daobao* [China high-tech industry herald], 16 December 1996.

28. Quoted in *Zhongguo Minban Keji Shiyejia Xiehui Huixun* [Newsletter of the China nongovernmental science and technology entrepreneurs association], April 1991.

29. Management Office of the Xi'an New and High-Technology Development Zone, "Jianshe Gao Xin Yuanqu, Zhenguang Shaanxi Jingji" [Build a new and high-technology park, invigorate the Shaanxi economy], in *Quanquo Kexue Jishu Dahui* [Documents from the national meeting], 154.

30. Management Office of the Xi'an New and High-Technology Development Zone, "Kua Shiji Chanye Jidi—Xi'an Gao Xin Jishu Chanye Kaifaqu" [The base of next century's industry—Xi'an new and high-technology industry development zone], in *Guojia Gao Xin Jishu Chanye Kaifaqu Suozai Shi Shi Zhang Zuo Tanhui: Wenjian Zailiao Huibian* [Collected documents from the meeting of mayors from cities with national level new and high-technology industry development zones] (Beijing: Guojia Kewei Huaju Jihua Bangongshi, 1996), 154.

31. Compare this to how Shanghai's *minying* enterprises were overlooked with that city's strategy of "grasping the big, releasing the small." "Shaanxi Sheng Renmin Zhengfu Guanyu Shishi 'Shaanxi Sheng Minying Keji Qiye Tiaoli' Jin Yi Bu Fazhan Minying Keji Qiye Shiye de Jueding" [Decision concerning the Shaanxi people's government implementing "regulations about Shaanxi Nongovernmental technology enterprises"], in *Keji Fagui Xuanbian* [Selected S&T laws and regulations] (Xi'an: Xi'an Kexue Jishu Weiyuanhui, 1996), 412.

THE ROLE OF THE GOVERNMENT:
FINANCIAL POLICIES IN GUANGZHOU

A conception of nongovernmental enterprises as more central to local development emerged in new policies both in Guangzhou and Xi'an in the early and mid 1990s. But until these changes occurred, policies dealing with technological innovation centered on each locality's primary economic concerns: attracting and retaining foreign investment in Guangzhou, integrating technological adjustment and protecting the position of SOEs in the public sector in Xi'an.

The Background to Financial Policies: Guangzhou

Paradoxically, *minying* enterprises in Guangzhou faced the same shortage of start-up capital that severely limited the growth of nongovernmental technology enterprises in other parts of the country. Guangdong's growth during the reform period was generally the story of increased control over rising local revenues, the inflow of large amounts of foreign investment, and encouraging private industry. So, if local leaders in Guangzhou had both the resources for and the interest in developing high technologies, why were *minying* enterprises starved for capital?

Part of the paradox stems from the fact that although other cities in the province were increasing their control over their own revenues, Guangzhou remained tightly controlled by the province. In 1979 the central government granted Guangdong the right to retain all revenues above RMB 1.2 billion for five years. Because it was not the ratio of remittances that were fixed but the actual amount, this fiscal regime was far more favorable to Guangdong than the contracts granted to other provinces.[32] Following the establishment of the fiscal contract system, the central government further outlined the "special policies and flexible measures" the province was to enjoy, and they included measures granting substantial autonomy in investment, foreign trade, and the management of commercial activity.[33]

Decentralization at the central level was reproduced at the provincial and subprovincial level. As Guangdong did to the center, Guangzhou now remitted a fixed percentage of revenue to the province annually and was understood to enjoy greater freedom to promote development as it saw fit. But throughout the reform period, Guangzhou's role as a "cash cow" for the province meant that it did not secure the increasing control over its budgets that Shenzhen did. In 1992, for example, Guangzhou's remittance

32. Vogel, *One Step Ahead in China.*

33. See Jia Hao and Lin Zhimin, ed., *Changing Central–Local Relations in China: Reform and State Capacity* (Boulder, Colo.: Westview Press, 1994), 213–14. The documents are: Central Document no. 50 (1979), Central Document no. 41 (1980), and Central Document no. 27 (1981).

to the center and province amounted to 90 percent of expenditures that year.[34]

The other part of the paradox can be explained by incentive structures. Guangzhou officials in the late 1980s and early 1990s invested where returns were highest. And during this time, investing in real estate and trading companies was much more profitable than putting money into high-risk research and development. Between 1991 and 1995 investment in industry in Guangdong province increased by 27.3 percent per annum; investments in real estate increased by 76.6 percent.[35] Shenzhen created incentives to redirect this money into high technology, offering preferential policies to real estate and trading companies who invested in high-tech products, and as a result Shenzhen was widely seen to be investing more in high-technology enterprises than Guangzhou.[36] Moreover, much of the growth in TVEs and private enterprises in Guangzhou was fueled by the pent-up demand for consumer products, including household electronics and appliances. Enterprises with any technological base found it easier to exploit these markets and satisfy consumer demand than develop high technology.[37]

Bank Lending: Guangzhou

Initial investment in Guangzhou *minying* enterprises came from either the supervisory agency or was self-raised by individual scientists among their family and friends.[38] In 1987 the city issued its "Decision on Using Guangzhou's Technological Strengths and Accelerating Horizontal Cooperation" that instructed the city's industrial and agricultural banks to issue loans to encourage enterprise R&D. The decision also provided preferential loans to enterprises that bought high-technology products from research institutes and gave tax breaks to SOEs that set up joint production entities with research units.[39] In the same year, the city issued its "Temporary Measure for S&T Lending," instructing the city Finance Bureau and the Guangzhou branch of the Bank of China to help with S&T loans.[40]

34. Peter Y. Cheung, "Guangdong's Advantage: Provincial Leadership and Strategy toward Resource Allocation since 1979," in *Provincial Strategies of Economic Reform in Post-Mao China: Leadership, Politics, and Implementation*, ed. Peter T. Y. Cheung, Jae Ho Chung, and Zhimin Lin (Armonk, N.Y.: M. E. Sharpe), 119.

35. Ibid., 133.

36. Shenzhen People's Government, "Fahui Zhengfu Zhudao Zuoyong, Cujin Gao Xin Jishu Chanye Quan Fangwei Fazhan" [Develop the government's leadership role, accelerate new and high-technology industries all-around development], in *Shi Zhang Zuo Tanhui: Wenjian Zailiao Huibian* [Collected documents from the meetings of mayors], 175–80.

37. Kuang, "Dali Fazhan Guangdong Sheng" [Energetically develop Guangdong province], 13.

38. "Guangzhou Chuxian Keyan Shengchan Yitihua de Shi Zhong Xingshi" [Circumstances for the appearance of joint science and production units], *Keji Ribao*, 7 February 1988.

39. "Guangzhou Wei Keji Ti Gaizhi Ding Peitao Zhengce" [Guangzhou enacts a set of policies to promote reform of the S&T system], *Keji Ribao*, 14 July 1987.

40. Ibid.

Even after *minying* enterprises became eligible for loans in 1985, high-technology enterprises were not particularly successful in securing them. In 1990, for instance, only 3.2 percent of R&D funds in high enterprises came from bank or government lending.[41] In contrast, local officials in Shenzhen implemented a more aggressive policy in support of bank lending to *minying* enterprises, acting as their advocate and guarantor for loans. In 1995, of 278 high-technology enterprises surveyed, 2.9 percent of operating capital came from the government, 18.2 percent from banks, and 78 percent was self-raised.

The case of Suntek Group (Xintai Jituan) exemplifies the problems faced by even the most successful nongovernmental enterprises in securing loans. Two researchers, Deng Longlong and Zhao Taisu, left the Micro-electronics Research Institute of Zhongshan University in 1986 to establish the company; start-up capital was entirely self-raised, and by 1995 the company had grown to over five hundred employees with assets of over RMB 250 million.[42] Deng Longlong was named a model worker, managers described the relationship between the company and the city government as good, and yet access to capital remained the biggest barrier to the enterprise's continued growth. In 1995 the company set up a joint venture with the local branch of the Ministry of Post and Telecommunications and the city government. To finance this project, Suntek eventually received a loan. But managers complained that they had to buy and develop a large office building as collateral for the banks; one manager noted "that building was a waste, and in the United States we would have received investment just with a good business plan."[43]

Government Investment: Guangzhou

In addition to expanding bank lending, the municipal government addressed capital scarcity by investing in research and development and creating specialized financial institutions. In both of these areas, government attention narrowed from the needs of technology development broadly defined to the needs of *minying* enterprises. The provincial Science and Technology Commission, for example, established the Newly Emerging Science and Technology Plan (*Xinxing Keji Jihua*) in 1987 to support research in information technologies, biotechnology, and new materials. In the same year, the commission also used an additional RMB 3 million to establish key labs in basic research in different parts of the province.[44] In 1994 the local government created specialized research labs in critical technologies

41. These are not broken up by enterprise type. Wu, "Guangzhou Gao Xin Jishu Chanye" [Countermeasures and evaluations], 28.

42. "Keji Qiangren: Deng Longlong" [Science and technology superman: Deng Longlong], *Dushi Ren* [Capital people] 10 (1997): 110.

43. Interview, G16, 6 July 1998.

44. "Keji Tizhi Gaige Dayou Gaotou" [Reforms of the S&T system], *Keji Ribao*, 5 April 1987.

like biotechnology and IT. And in 1995 the city allocated RMB 20 million for renovation and improvement of equipment in science research institutes.[45] Yet Guangzhou's ability to support these labs seems doubtful given the percentage of remittances handed over to the provincial government. Moreover, Guangzhou's S&T burden was higher than other cities; reportedly 70 percent of government-funded R&D in the city, including research supposedly undertaken by central institutes, was funded locally.[46]

The first specialized financial institution in the city was the Guangzhou Science Investment Corporation. Founded in 1987, the corporation was state-owned with support from the Guangzhou Science and Technology Commission, city finance bureau, and Guangdong Investment and Trust Corporation (GITIC); the corporation began with RMB 115 million to disperse to high-tech enterprises of all property types.[47] In 1991 the province and the city both established S&T Development Funds with more than RMB 300 million.[48] And in 1994, the provincial Science and Technology Commission (STC), provincial development bank, and a U.S. venture capital company, Pacific Venture Capital, founded the Guangdong Pacific Pioneer Limited Corporation (Guangdong Taiping Yang Chuangye Youxian Gongsi) with over RMB 100 million.[49]

In late 1994, the city conducted a study of nongovernmental enterprises and the problems they faced in development. According to the report, most enterprises had problems securing access to capital and land. In order to address capital scarcity, the STC created a Science and Technology Credit Association (Keji Xinyong She) to help small enterprises. Also, with the help of a Hong Kong businessman who put up half of the money, the Science and Technology Commission established the Guangzhou Technology Progress Fund (Keji Jinbu Jijinhui). Finally to prevent enterprises like Suntek from unnecessarily building office towers, the Science and Technology Commission created a loan guarantee and risk fund of RMB 10 million for nongovernmental enterprises.[50]

In April 2000 Guangzhou deputy mayor Lin Yuanhe announced that

45. "Guangzhou Increases Science, Technology Funding," *Xinhua*, 21 November 1995, in FBIS-CHI, 22 November 1995.

46. "General Observations on the Reform Process," in *A Decade of Reform: Science and Technology Policy in China* (Ottawa, Canada: International Development and Research Center, 1997). Online at http://www.idrc.ca/books/815/chap04, accessed September 1999.

47. "Guangzhou Keji Touzi Gongci Chengli" [Guangzhou science and technology investment company established], *Keji Ribao*, 7 March 1987. GITIC was declared bankrupt and it seems to have not done very well with technology investments. Guangzhou entrepreneurs questioned whether the government could ever efficiently funnel money to high technology. Interview, G13, 3 July 1998.

48. Guangdong Science and Technology Commission, "Jianli he Jianquan Keji Touru Xitong, Jiasu Keji he Jingji de Fazhan" [Build and perfect the S&T investment system, speed S&T and economic development], in *Quanguo Kexue Jishu Dahui*, 128.

49. Ibid.

50. Ibid.

the city would establish a venture fund and a venture investment company to develop high-tech industrial projects.[51] Lin pledged that Guangzhou would attempt to raise RMB 1 billion for the fund in three years. The city government has also announced its plans to invest RMB 10 million in ten software companies and fifteen software development projects while Tianhe Science Park set aside another RMB 30 million for software development.[52]

Foreign Direct Investment: Guangzhou

Guangzhou adopted a funding strategy for technology highly reliant on foreign direct investment and joint ventures. In order to attract this capital, foreign enterprises in Guangzhou were offered a wide range of preferential policies. The terms of these policies were even more generous for those enterprises willing to locate in Tianhe Science Park and included exemption from income taxes for two years, tax breaks of 50 percent for up to six years, a return of 40 percent of paid income tax for reinvesting profits, and the ability to participate in Torch and other state-funded technological plans.[53] By comparison domestic enterprises designated to be involved in "new and high technology" by the city Science and Technology Commission received an income tax rate of 15 percent, and tax exemptions for only two years.[54]

After China's entry into the World Trade Organization, Guangzhou expects to take the lead in opening even further to foreign ventures. In November 2000 Weng Wenxiang, deputy director of Guangzhou's Foreign Economic and Trade Commission, announced that the city would make greater policy adjustments in order to attract foreign direct investment (FDI). These adjustments include further import tariff reductions, narrowed use of active quotas and import licenses, cancellation of the foreign exchange balance system, and greater access to domestic markets.[55]

By 1995 Tianhe was the base of over $100 million in investments. But domestic analysts have grown increasingly critical of relying solely on FDI.

51. "Deputy Mayor Says Guangzhou to Launch High-Tech Fund," *Xinhua*, 6 April 2000, in FBIS-CHI, 7 April 2000.

52. "Guangzhou Ruanjian Yuan Zhen Zhui Beijing Zhongguancun" [Guangzhou's software park is really chasing Beijing Zhongguancun], *Guangzhou Ribao* [Guangzhou daily], 12 September 1999.

53. Exemptions from income tax varied depending on the number of years of operation and whether or not the foreign-funded enterprise was designated as a "new and high-technology enterprise." If so, these enterprises enjoyed a two-year complete exemption from income tax, followed by an additional six years of half exemption. In addition, if these enterprises used profits to establish other "technologically advanced" enterprises for a period of no less than five years, the income tax levied on the invested amount was to be returned in total. See "Preferential Policies for Foreign Joint Ventures," in *Investment Guide: Guangzhou Development Zone for New and High-Technology Industries Tianhe STC-Tech Park* (Guangzhou, n.d.).

54. "Preferential Policies for Internal Enterprises," in *Investment Guide*.

55. Zong Xin, "After China's Entry into the WTO, Guangzhou Will Take a Lead in Opening Seven Major Trade Sectors," *Yangcheng Wanbao*, 27 November 2000, in FBIS-CHI, 27 November 2000.

They argue that high-technology development based on foreign investment was a more complicated process than just attracting investors from
abroad. The process also entailed the absorption, diffusion, and innovation
of new technologies, and these have not been occurring in Guangzhou.[56]
FDI has been mainly concentrated in assembly operations and low-tech
products. For instance, the Guangzhou Science and Technology Commission declared in 1992 that only forty, or 5 percent, of Hong Kong invested
enterprises were actually involved in high-technology processes or products. And of RMB 213 million in products exported to Hong Kong, only
RMB 28 million (13 percent) were considered high technology.[57]

Two other problems prevented the diffusion of technological capabilities to Guangzhou enterprises. First, the property rights structures of joint
ventures left control of technology decisions in the hands of the parent
company. Foreign managers were more interested in selling products in
the domestic market than modifying technology for local needs. Second,
China has traditionally underestimated the difficulty of and thus under
funded the absorption of new technologies. According to a 1994 report by
the State Science and Technology Commission, for every yuan spent to introduce new technology, China spent only 0.9 yuan for the digestion and
assimilation of that technology. In Japan and Korea the expenses for digestion and assimilation were close to ten yuan for every one yuan spent.[58]

The more general problem was FDI could not substitute for indigenous
technological development in Guangzhou.[59] Foreign enterprises did not
transfer higher value-added activities to their local partners since they were
not assured that important inputs were available locally. The ability to find
these inputs was directly linked to the general environment in Guangdong
and included local subcontractors and the protection of intellectual property rights. Although the investing enterprise may have had a limited impact through training and other projects, the main responsibility for raising the overall technological level of the economy still rested with the
Guangzhou local government.

FINANCIAL POLICIES IN XI'AN

Minying entrepreneurs in Xi'an also had to battle to receive funding, but
they were operating in an environment of overall scarcity. In Guangdong,

56. Zheng Yinglong, "Guangzhou Gao Xin Jishu Chanye de Fazhan yu Waixiang Xing Xuanze" [Guangzhou's new and high-technology industry development and its decision to orient
to the outside], *Keji Guanli Yanjiu* [Research in scientific management] 4 (1994): 29–34.

57. Both set of figures from Liang Jianing, "Dui Sui Gang Keji Hezuo de Rougan Sikao"
[Some thoughts on S&T cooperation between Guangzhou and Hong Kong], *Zhongguo Keji Luntan* [China S&T forum] 3 (1994): 54–56.

58. Cited in *A Decade of Reform*.

59. This paragraph draws heavily from Jean Francois Huchet, "The China Circle and Technological Development in the Chinese Electronics Industry," in *The China Circle*, 272.

provincial leaders learned to cope with retaining increased revenues generated by a rapidly expanding economy. Money did not go to nongovernmental enterprises because there were so many other places for it to go. In interior provinces like Shaanxi, the problem was not enough money; provincial leaders were engaged in a constant struggle to raise revenue. Since 1979 Shaanxi has continued to run increasing debts and became a deficit province during the reforms.

In the prereform era, Shaanxi's growth and development relied on transfers from the central government; central government investment in the province was higher than the national average, and the central government established and controlled the industries that dominated the local economy. Average central contribution to total investment was 79 percent.[60] The coastal development strategy sharply curtailed central government investment, and Shaanxi's share of national investment fell. Tax reforms did not improve the province's position. With a high proportion of enterprises operating at a loss and a large number of state-owned enterprises submitting their revenues directly to the center, efforts to establish a tax-based fiscal system did little to increase access to revenues for provincial leaders.

Within the province, Xi'an was the major source of revenue, providing almost one-third of provincial totals. During the reform period, Xi'an contributed 50 to 70 percent of its earning to the province. In the 1980s, the provincial government encouraged lower levels to establish their own fiscal administrative offices, giving these units greater responsibility and control over resources. These changes, however, did not dramatically improve Xi'an's situation since a high percentage of SOEs operated at a loss. In 1994, 63 percent of state-owned enterprises were in the red.[61]

Because of the inefficiency of Shaanxi SOEs (as well as their isolation), the province had difficulty competing with coastal provinces for domestic or foreign capital. In fact, there was a substantial outflow of the province's capital to coastal areas. Local banks lent their money to coastal customers, individuals invested on the Shenzhen or Shanghai stock exchanges, and collectives and individuals bought property in Guangdong or Fujian. Capital outflows exceeded RMB 2 billion in the first half of 1993 alone.[62]

In January 2000 the State Council began addressing the growing disparity between the hinterland and coastal provinces through a western development strategy. The portion earmarked for western development in the central government's fiscal budget was set to rise from one third to about two fifths. The central government also expanded the areas open to foreign investors, broadening the scope of overseas investment in western

60. Zhao Bingzhang and Zhang Baotang, *Shaanxi Jingji Fazhan Zhanlue Zonglun* [Survey of Shaanxi's economic development strategy] (Xi'an: Sanqin Chubanshe, 1988), 327.

61. Cited in Lane, "One Step Behind," 229.

62. Ibid.

provinces like Shaanxi and adopting more flexible policies in foreign trade. The project included increased investment in science and education as well as construction of roads, airports, railroads, and a $14 billion, twenty-five-hundred-mile pipeline linking the West's natural gas fields to the energy-hungry East.

Bank Lending, FDI, and Returned Funds: Xi'an

Given that money was flying out of Xi'an during the 1980s and 1990s, it is easy to see why *minying* enterprises had a difficult time securing access to funds from either banks or the local government. The municipal government's focus on SOEs (and their increasing insolvency) only heightened this difficulty. *Minying* and technology enterprises in general had difficulty securing loans without some type of government support, which was late to come and often incomplete. Most banks in Xi'an required immovable collateral to issue loans to small enterprises and they preferred real estate.

In 1994 the Xi'an High-Technology Zone established a loan guarantee program.[63] Within the zone, the small business incubator controlled a fund of about RMB 40 million to invest.[64] In 1996 the city government and the STC established an investment center for small- and medium-sized enterprises, issuing loans totaling RMB 5.8 million to thirty-one companies. The finance bureau was told to increase the size of loans and credit support to *minying* enterprises the same year. Also Xi'an established a Nongovernmental Technology Enterprise Credit Guarantee Company (Minying Keji Qiye Xindai Danbao Gongsi).[65] The first scientific and technological venture investment company in western China, the Xi'an New and High-Tech Risk Investment Company, was founded in August 1999.[66] With capital of RMB 50 million, the company hoped to overcome "capital bottlenecks" by funding high-technology industries. The local government will defray 80 percent of the risk incurred by the company. In 2000 the zone in cooperation with the Xi'an Science and Technology Commission set up a small fund for young people with S&T backgrounds to help them establish their

---

63. Xi Ke, "Zujian Daikan Danbao Jigou de Shijian yu Tihui" [The practice and experience of establishing a loan guarantee organization], *Zhongguo Gao Xin Jishu Chanye Daobao* [China new and high-technology industry herald], 23 March 1997.

64. "Xi'an Gao Xin Qu Zhaoshang Yinzi Xin Tupu" [Xi'an high-technology zone makes new breakthroughs in attracting foreign investment], *Zhongguo Gao Xin Jishu Chanye Daobao*, 28 March 1997.

65. See "Guanyu Zengjia Xianshi Keji Daikuan E Du de Shishi Yijian" [Implementing the decision to increase Xian's S&T loans], and "Shaanxi Sheng Renmin Zhengfu Guanyu Shishi 'Shaanxisheng Minying Keji Qiye Tiaoli' Jin Yi Bu Fazhan Minying Keji Shiye de Jueding" [Decision concerning the Shaanxi people's government implementing regulations about Shaanxi nongovernmental S&T enterprises], in *Keji Fagui Xuanbian* [Selected S&T loans and regulations], 354.

66. "First Company Specializing in S&T Venture Capital Founded," *Xinhua*, 19 August 1999, in FBIS-CHI, 19 August 1999.

own companies. The average investment was around RMB 500,000.[67] Later in the same year, the Xi'an New and High-Technology Development Zone began allowing investors to establish venture capital investment companies. The companies take the form of limited liability companies, provided they have a total investment of no less than RMB 500,000 and no more than fifty shareholders. In other parts of China, the minimum threshold for venture capital investment is typically set higher, at RMB 10 million for domestic investors, RMB 248 million for foreigners.[68]

Rates of foreign direct investment picked up after 1992 when the central government designated Xi'an as an "internal open city" and allowed it to offer the same benefits to foreign investors as coastal areas. In 1992, the entire province of Shaanxi only had 78 foreign-funded ventures with total output of RMB 900 million; by 2000 the number of foreign invested enterprises exceeded 1,800.[69] As in Guangzhou, the high-technology development zone provided preferential treatment for joint ventures, including tax breaks of up to eight years and low (and at times free) rents.[70] By 1996 a total of 256 joint ventures with foreign investment of $350 million were located in the Xi'an high-technology zone.[71] Furthermore, 70 large- and medium-sized enterprises, 30 research institutes, and 20 universities and other higher education units were chosen by the local government to develop cooperative agreements with foreign partners.[72]

A limited amount of funds returned to Xi'an from coastal provinces. Some former residents, like the Jiaotong University student who saved RMB 100,000 working in Shenzhen and returned to open his own CD-ROM development company, believed that Xi'an's technological base made it a more appropriate location to start a company than Guangdong.[73] Jiaotong professor Hou Yibin established a branch of his KaiTe Technology Group (a developer of Chinese-language software) in the Xi'an Technology Development Zone.

Financial Policies and Xi'an Minying Enterprises

The case of Haixing (Seastar), Xi'an's most famous and arguably most successful *minying* enterprise, illustrates how capital restrictions affected enterprise behavior. In 1986 Rong Hai left his research institute at Xi'an Jiaotong University with several other professors to set up a computer network company. The enterprise registered as a collective enterprise and

67. "Xi'an Qingnian Keji Rencai Chuangye Jihua Shoubi Fuchi Xiangmu Queding" [First projects in Xi'an S&T youth pioneers plan receive support], *Keji Ribao*, 24 February 2000.
68. "Threshold for Capital Investment Lowered in Xi'an," *China Online*, 10 January 2001.
69. "Shaanxi Governor Discusses Economic Strategy," *Shaanxi Ribao*, 15 May 1996, in FBIS-CHI, 15 May 1996.
70. Interview, X2, 22 July 1998.
71. Management Office of Xi'an Development Zone for High-Technology Industries, *The State Level Xi'an Development Zone for High-Tech Industries* (Xi'an: n.p., 1997).
72. "Xi'an Gao Xin Qu Zhaoshang Yinzi Xin Tupu" [Xi'an high-technology zone].
73. Interview, X5, 23 July 1998.

bought equipment and rented office space with the initial start-up capital of RMB 30,000 raised by Rong Hai.[74] By 1996 Haixing's total assets amounted to RMB 260 million.[75] The company's early success was almost entirely dependent on the shrewd use of representative agreements. Haixing became Compaq's representative for Northwest China in 1991; during the next two years the company aggressively built a distribution network, passing RMB 100 million in sales.[76] In 1993, when Compaq dominated the domestic market, Haixing became Compaq's representative for all of China. In the following years, Haixing signed agreements with IBM, Intel, Hewlett-Packard, Canon, Fujitsu, and other foreign companies. These agreements not only allowed Haixing to study how foreign companies entered new markets, but also created a steady capital flow.

The company used much of this capital to start an aggressive strategy of diversification. Computers were to remain the core of Haixing's business, and in 1996 it introduced the Haixing Dragon PC, selling twenty thousand units in the first half of the year. By 1997 the scope of business had broadened to involve supermarket chains, soft drinks, tropical agriculture, film and TV production, and real estate. In Xi'an, down the street from the offices of the provincial government, Haixing built a RMB 100 million "intelligent" office complex. Rong Hai was elected to the Ninth National People's Congress, and Haixing was selected as one of the key "giant" enterprises to be specially supported by the provincial government.

The size and success of Haixing provoked a wide range of reactions from nongovernmental entrepreneurs in Xi'an.[77] Many pointed out that the company's original growth had little to do with technological savvy or capability; growth was the result of Rong Hai's skill in building a distribution network for Compaq, not the quality of the original software he had developed. All argued that Haixing depended on government support. Officials at the enterprise did not deny close relations with the provincial and local government, or that those connections were important; "Without government support, nongovernmental enterprises could not make it."[78] Both levels of government now introduced Haixing to potential foreign partners. Moreover, the local government itself became an important partner. Haixing, the Xi'an municipal government, and a local SOE jointly invested

74. Rong Hai left Jiaotong in 1986 and founded the original start-up that would eventually become Haixing in 1988. See "Rong Hai, Banian Fangfei" [Rong Hai: Eight years in flight], in *Mingying Keji Qiye de Fazhan* [The development of nongovernmental technology enterprises], ed. Wang Jianhua (Beijing: Jingji Kexue Chubanshe, 1996), 232–43.

75. Haixing Science and Technology Group, company prospectus, n.d.

76. Yao Jiang, "Haixing: Yige Xibu de Chuanqi Gushi" [Seastar: A western legend], *Guoji Dianzi Bao* [International electronics journal], 24 January 1994.

77. Many of them angry. In response to a question about the Dragon PC, one entrepreneur said it was a piece of "garbage" and doubted they had sold one, much less 20,000. Another called Rong Hai more politician (*zhengke*) than entrepreneur (*qiyejia*), but then argued that you had to be to be successful. Interview, X4, 23 July 1998.

78. Interview, X8, 27 July 1998.

RMB 110 million in a computer factory; Haixing was responsible for 60 percent of the investment, the rest equally divided between the two other partners. In 1997, Haixing, the General Northwestern Electric Power Development Company, and seven others established a large information industrial group. Shaanxi governor Cheng Andong described Haixing Modern Technology as a "masterpiece" in the province's effort in establishing a modern enterprise system.[79]

These close connections led one analyst to argue that Haixing was no longer a nongovernmental enterprise. Although there had been no actual change of ownership, the local government had invested so much political prestige in the enterprise and was now willing to bail it out with endless loans that it could operate with the soft-budget constraints of a SOE.[80] This relationship may limit Haixing's independence, but it was a trade-off some nongovernmental enterprises were willing to make. As one Xi'an entrepreneur exclaimed, "I wish the local government would pay attention to me even if that means they are going to interfere with internal management. At least I would be able to get a loan."[81]

DEFINING PROPERTY RIGHTS

Guangzhou and Xi'an moved relatively late to define *minying* as a separate form of technology organization, but in different ways and for different reasons. In Guangzhou, local officials generally subsumed nongovernmental into private. This meant rapid growth in small-scale, nonstate enterprises, but also resulted in a lack of indigenous innovation capability. In particular, the Guangzhou government was slow to support stock systems for technology, an important tool in attracting and keeping talented scientists. Property rights regulations for *minying* enterprises in Xi'an during the 1980s were a subset of the conservative attitude local leaders adopted to reform overall. Officials only slowly recognized and regulated nongovernmental enterprises; their overriding objective was to protect the dominant position of SOEs in the local economy. During the early 1990s this attitude began to change and the local government tried to convert small and collective enterprises to more private forms of ownership.

Regulating the First Minying *Enterprises: Guangzhou*
The first regulations about S&T personnel and enterprises in Guangdong appear to have concerned town and village enterprises. Starting in 1983, TVEs began receiving preferential treatment and were encouraged to hire S&T personnel; many scientists went to work in rural areas as "Sat-

---

79. "Large Information Industrial Group Formed in Shaanxi," *Xinhua*, 16 December 1997, in FBIS-CHI, 17 December 1997.
80. Interview, X6, 24 July 1998.
81. Interview, X4, 23 July 1998.

urday engineers" (*xingqi liu gongchengshi*), or part-time consultants.[82] Scientists and S&T units did not begin setting up their own *minying* enterprises until after 1984. Research institutes and SOEs set up small enterprises or service companies, research institutes became enterprises themselves, or research units founded alliances with collectives or SOEs and the research institute became the R&D unit of the newly formed enterprise.[83]

In 1985 the Guangzhou Science and Technology Commission issued a decision regulating *minying* enterprises, describing how these enterprises should be registered and what should be done to encourage them.[84] This decision appears to have been in response to the national-level 1985 decision and was not followed by the issuance of further regulations at the local level. Moreover, the Science and Technology Commission was unable to convince other departments of the need for the category of nongovernmental science enterprise. At a provincial S&T development meeting in Zhuhai in 1997, representatives of the Industrial and Commercial Bureau argued that the term should be abolished since they were the ones who applied it when registering new enterprises and they had no idea what it meant.[85]

In a type of regulation not observed in Xi'an, Beijing, or Shanghai, Guangzhou legalized the position of S&T personnel as "middlemen." According to one newspaper report, these people, who in the past would have been called brokers (*qianke*), were selling technology products from their own or other research units to production enterprises.[86] Since this behavior was at best considered part of the "individual" (*geti*) economy and at worse illegal, the Guangzhou STC and the Industrial and Commercial Bureau created a ten-day course to license official S&T middlemen. The existence of the course and the license gives weight to one Xi'an entrepreneur's claim that "*minying* activity in Guangzhou is not developing new technologies. It is all buying and selling."[87]

Property Rights Regulations and Enterprise Behavior: Guangzhou
This pattern of regulation created at least four problems for Guangzhou entrepreneurs. First, as the 1997 S&T meeting in Zhuhai demonstrated, not clearly defining nongovernmental enterprises created confusion. In one case, seven or eight members of the provincial Anti-smuggling Inspection Bureau entered the office of a nongovernmental software design company. Inspectors discovered that some of the design software had been im-

82. "Keji Tizhi Gaige Dayou Gaotou" [Reforms of the S&T system], *Keji Ribao*, 5 April 1987.
83. "Guangzhou Chuxian Keyan Shengchan Yitihua de Shi zhong Xingshi" [Circumstances for the appearance of joint science and production units], *Keji Ribao*, 7 February 1988.
84. The information in this paragraph is from interview, G5, 12 November 1996.
85. Ibid.
86. "Guangzhou Shi Kewei You Jihua Peixun Jishu Jingji Ren" [Guangzhou STC has plan to train economic and technological personnel], *Keji Ribao*, 21 February 1988.
87. Interview, X3, 23 July 1998.

ported from abroad, and since the company did not officially have an import license, the bureau froze company funds totaling RMB 400,000. In fact, the company was legally allowed to use the imported technology under a provision called "second-use technology development." Members of the Anti-smuggling Bureau explained that they knew nothing about the technology, nothing about the provision, and nothing, for that matter, about nongovernmental technology enterprises.[88]

Second, regulations like the 1994 "Decision on Private Enterprises Involved in S&T" made it too easy for foreign firms to register as private enterprises in Guangdong. When private enterprises first emerged in Guangzhou in the early and mid 1980s, they registered as collective; private ownership, even in Guangzhou, was considered too politically sensitive. Suntek, for example, registered as collective even though the company was in fact completely private. Much of Guangzhou's early work in property rights was directed at clarifying the property rights of collective enterprises. As one official at the Tianhe High-Technology Zone explained, "We help collectives become what they really are. If a company is private they should just say they are private."[89]

Yet, as far as nongovernmental enterprises were concerned, local officials may have opened up the meaning of private too far. As one professor involved in a start-up company complained, Japanese and Taiwanese firms could set up a representative office in Hong Kong, find a mainland Chinese manager, and set up shop in Guangzhou. The ease of registering in Guangzhou as a private enterprise meant foreign firms avoided the bureaucracy involved in setting up a joint venture; three or four days after arriving they were in business. Local enterprises could not compete.[90]

Third, Guangzhou, content to see most enterprises as private, did little to promote special policies specifically supportive of nongovernmental technology enterprises. This has created an unexpected pattern where enterprises originally registered as private or collective actually became partly state owned. An example can be given from a biotechnology company, but the case is still illustrative. Involved in the development of urokinase, an enterprise registered in Nanjing as an individual enterprise; seven friends raised the start-up capital of RMB 60,000. After moving to Guangzhou, the company eventually created a joint venture with two SOEs, one from Guangzhou, the other Shenzhen. The two SOEs controlled 60 percent of the new company, 40 percent remained in the hands of the original biotechnology enterprise. In the words of the founder, "Being an SOE means that it is easier to get loans and easier to get people to cooperate with us. People discriminate against *minying* enterprises, but the state protects its own."[91]

88. "Zhe Yang 'Guan Si' Ruhe Liaojie" [This is how to understand 'lawsuit'], *Keji Ribao*, 12 September 1987.

89. Interview, G17, 8 July 1998.

90. Interview, G10, 29 June 1998.

91. Interview, G13, 3 July 1998.

Finally and closely related to the third point, the belated development of internal stock systems for *minying* enterprises reduced Guangzhou's ability to compete with Zhuhai and Shenzhen for talented scientists. Both cities initiated very aggressive campaigns to lure talented S&T personnel to local enterprises, providing benefits like houses, high salaries, and cars.[92] But the ability of enterprises in these cities to offer scientists technology stocks was equally important in attracting S&T talent. Guangzhou only began experimenting with this system in 1998; Shenzhen had adopted it in 1994.

Regulating Property Rights: Xi'an

As with other reform policies, provincial and municipal officials in Xi'an reacted slowly to the emergence of nongovernmental enterprises and failed to support the institutions that preceded the emergence of competitive *minying* enterprises. In 1984, for instance, a number of industrial enterprises and S&T institutes established joint research-production units to help find commercial uses for some high-technology processes. The author of a report on these units noted, however, that Xi'an lagged behind other cities in both the quantity (thirty-five) and quality of these units and listed two main reasons for this. First, managers of SOEs in Xi'an were satisfied with their current production technologies and uninterested in the use of new technology. Second, the joint units lacked organized support from the local government.[93]

The author of a report on enterprises set up by universities in Xi'an came to a similar conclusion. By 1990 universities in the city had established 160 independent enterprises, but not one could be called high technology. Why? Universities needed development capital and they lacked entrepreneurs who could bring new products to market. Also the city had again failed to provide organized support for the enterprises.[94]

During the early 1990s, local officials became much more proactive about S&T development, and by 1995 the web of policies, regulations, and laws surrounding nongovernmental enterprises was more comprehensive. In 1990, the *Minying* Entrepreneurs Association reported there were 352 *minying* enterprises in Xi'an, 313 (89 percent) of which were collective.[95] During the next five years, provincial and municipal officials issued regulations and rules designed both to increase the number of enterprises and to convert collective enterprises to more clear ownership structures.

In 1992 Shaanxi issued the "Decision Regarding Management of State

92. James Tyson, "Southern China City Offers Kingly Perks to Lure Scientists," *Christian Science Monitor,* 21 February 1992.

93. Ai Zhangan, "Xi'an Shi Ke Yan Shengchan Lianheti Fazhan Qingkuang de Diaocha" [Report on the development of Xian's joint production research units], *Kexue Jingji Shehui* [Science, economics, society] 4 (1985): 306–8.

94. Liang Keyin, "Fahui Gao Xiao Keji Youshi, Fazhan Gao Xin Jishu Chanye" [Take advantage of the strengths of institutes of higher education, develop new and high-technology industry], *Kexue Jingji Shehui* 2 (1994): 20–22.

95. *Zhongguo Minban Keji Shiyejia Xiehui Huixun* [Newsletter of the China nongovernmental science and technology entrepreneurs association], 24 November 1990.

Assets in Technology Development Enterprises Established by Research Institutes" seeking to clarify the relationship between technology enterprises and their public-sector supervisory agencies, to protect the rights of state and other investors, and to encourage the creation of enterprise stock systems.[96] The statute declared that all new enterprises created after 1992 must register with clear property rights. Enterprises must also establish a board of directors, have a separate office to handle ownership questions, and create other "visible property rights organizations." At the other end of the relationship, research institutes must appoint an individual, at the director level, to be responsible for property rights restructuring.

After a series of meetings with *minying* entrepreneurs in 1992 and 1993, the provincial and municipal Science and Technology Commissions issued four decisions concerning *minying* enterprises. Among the four measures the most important was "Measure for Clarifying Nongovernmental Technology Enterprise Property Rights Relations."[97] This measure sought to clarify what *minying* actually meant as a property rights relationship. It also adopted the use of stock systems, including cooperative stocks (*gufen hezuo*), to further rationalize property relations. In creating stock systems, those involved were to address four questions—who invests, who owns, who collects income, and who bears the risk—and so invoked the concerns of the founder, managers, investors, and employees of the technology enterprise. The measure made the management office of the high-technology zone responsible for applying the decree, and in 1994 the provincial government chose five companies in the zone to experiment further with stock reform.

Property Rights Regulations and Enterprises Behavior: Xi'an
The implementation of stock systems among private, collective, or partially state-owned nongovernmental enterprises was not completely uneventful. Haixing was one of the first *minying* enterprises to implement an internal stock system. After the company's initial growth spurt, some of the original managers wanted to leave the company. With the departure of some of these scientists, the company was reorganized. Rong Hai declared that the only way for the company to insure future growth was for "everybody to have a share."[98] The company created a two-tiered internal system,

96. "Guanyu Keyan Jigou Xingban Keji Kaifaqu Qiye Guo You Zhi Chan Zhixing Guiding" [Decision regarding management of state assets in technology development enterprises established by research institutes], *Keji Fagui Xuanbian*, 4 October 1994, 424.

97. "Guanyu Lishun Minying Keji Qiye Chanquan Guanxi de Banfa" [Measure for clarifying nongovernmental technology enterprise property rights relations]. The other three were "Shaanxi Sheng Minban Keji Jigou Guanli Zanxing Guding," "Guanyu Cujin Minban Jigou Shiye Fazhan de Rougan Guiding," and "Guanyu Shenhua Keji Tizhi Gaige Dali Fazhan Minying Keji Qiye de Yijian." All collected in *Keji Fagui Xuanbian*.

98. "Everybody" really seems to apply to everybody in management. See "Minying Qiye Ruhe Qude Chenggong" [How a nongovernmental enterprise obtained success], *Zhongguo Jingying Bao* [Chinese management herald], 1 July 1994; and interview, X8, 27 July 1998.

providing salaries plus share holdings for the top S&T personnel and department managers, and a base salary and social welfare benefits for all other employees.

But by 1994 problems had already emerged with this system. Haixing had issued a limited number of stocks and had no way of redistributing existing shares or issuing new ones. New managers and scientists who arrived at Haixing later held no shares in the enterprise; some older managers who were now less engaged in the company had shares that reflected past responsibilities. Moreover, this system, while addressing the needs of Haixing, was in no way widely understood and so made cooperation with international corporations difficult.[99] In 1997, the company began experimenting with a new internal system that was to cover all employees and insure that profits and individual benefits increased at the same speed.

When public assets were involved, the efficacy of property rights reform depended heavily on the attitude of the supervisory agency. Here an example can be drawn from a company linked to a Xi'an university that developed new production technologies and the software that runs the machines involved in this process.[100] Two professors, who discovered the process while working at the university, founded a separate enterprise in 1990. The company consisted of three parts: a factory and production unit located at the university, a management and enterprise wing located in the Xi'an High-Technology Zone, and a research institute also located at the university.

It was over the position of the research institute that most conflict occurred. The institute trained masters and Ph.D. students who worked one-fourth of the time in the production or management enterprises, three-fourths of the time on their degrees; the institute received no direct funds from the university and instead was supported by profits from the enterprise. Even though the two professors argue that it did not exist before they founded their company, university officials considered the research institute as part of the university and annually demanded a portion of the profits of the enterprise. The percentage varied by year, but typically the company retained 15 percent and handed 85 percent over to the university. Of that 85 percent, the university reinvested 30 percent back into the research institute. Finally, the research institute was allowed to use up to 60 percent of the reinvested portion to distribute within itself as bonuses or salaries based on performance.

Administrative and contractual ties further complicated matters. The enterprise was chosen in 1994 by the State Science and Technology Commission as part of a national technological dissemination project (*Guojia Jishu Chengguo Zhongdian Jihua*) and received a RMB 500,000 investment

99. See "Xi'an Haixing Keji Jituan Zhua Jiyu Chuang Jiaji" [Xi'an Seastar Group seizes opportunity and has great results], *Zhongguo Gao Xin Jishu Chanye Daobao*, 11 June 1997.
100. Interview, X1, 21 July 1998.

from the central government. Relations with the management of the high-technology zone were extremely close, and officials in the zone pushed the enterprise to become more independent from the university. Finally, the enterprise licensed out its technology to SOEs that have become competitors, but who also act as partners when the enterprise has to contract out for large orders.

This case illustrates that a strong relationship exists between local government policy, innovative reform, and entrepreneurial management. Legend and CAS had a relationship as complex as the one between the university and the enterprise in Xi'an. But managers in both organizations were interested in clarifying those relations and could look to the local government to facilitate the process. The Beijing local government was already engaged in researching and promoting property rights reforms among nongovernmental enterprises. In those instances where the mind-set of planned economy dominates, like Xi'an, reform remained slow. Even if the Xi'an university had been more interested in clarifying property rights, the local government would not be prepared to serve in the same role as officials did in Beijing.

GOVERNMENT SUPERVISION

As they did with other policy spheres, Guangzhou and Xi'an adopted different strategies of supervising *minying* enterprises and organizing science parks. Like Shanghai, Guangzhou approached the zones as part of the local economy, not as concentrated areas of innovation separate from other developments. Moreover, it focused on attracting foreign investment as much as developing nongovernmental enterprises. In contrast, Xi'an adopted a strategy more like Beijing's, focusing its technological resources and the administrative powers to manage them into one area. Far removed from the coast, Xi'an, while certainly interested in drawing foreign investors, concentrated on local innovation.

Tianhe High-Technology Development Zone
In Guangzhou, technology zones in the city underwent several changes in management and administrative structures from 1991 to 1998.[101] The form finally adopted resembled Shanghai's "one zone, many parks" (*yi qu, duo yuan*). Like in Shanghai, city government control over the zones was extremely decentralized. All zones in the city were part of the Guangzhou High-Technology Development Zone (Guangzhou Gao Xin Jishu Chanye Kaifaqu) under the control of the city government. The municipal government also created an Office of the Management Commission of the

101. Section on high-technology zones in Guangzhou draws from interviews G3, 11 November 1996, and G17, 7 July 1998.

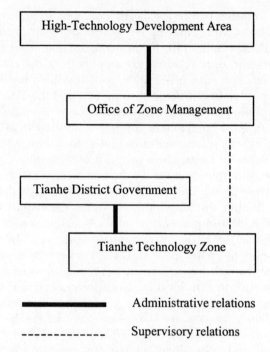

Figure 5.1. Guangzhou High-Technology Development Zone

Guangzhou High-Technology Development Zone with nominal supervisory power over the city's zone. Unlike in Shanghai, where the zones were connected to commissions within the city government, the zones in Guangzhou had administrative ties to the district governments in which they were located

A decentralized system lessened the burden on the city government. Each zone secured its own loans to develop new projects, and in the end was responsible for its own losses. But Guangzhou's decentralized system of management made it difficult for the city government to coordinate what all the zones were doing. Officials with ties to district governments met attempts by the Guangzhou STC to direct the internal management of the zones or coordinate action across administrative hierarchies with some resistance.

The Guangzhou city government established Tianhe in 1991, motivated in part by the hope that government policy could help move technological products more rapidly into the market. The Tianhe district government and the Guangzhou STC set up the original zone; each unit invested RMB 300,000 and staffed the management office with four cadres. The initial investment was used to build the administrative offices and clear land for factory sites. At the end of 1991, the city government invested another RMB 200,000 followed by investments of RMB 300,000 in 1992 and 1993 to keep the park running before it had attracted tenants and collected management fees, taxes, and rent. The city government also secured a low interest loan of RMB 1 million for further construction. After this initial period of support, the zone received no support from the city government and relied on taxes on enterprises in the zone for operating expenses.

Tianhe adopted a dual strategy of using FDI and the Torch Plan to develop high-technology industries. Preferential tax policies, low land prices, and promises of local government support were used to attract multinationals willing to set up joint ventures and share advanced technologies. Foreign enterprises in Guangzhou proper were exempted from local income, land use, and housing taxes for the first three years after their establishment and from paying the local government housing subsidies for Chinese workers. Within the zone, joint ventures that could meet criteria about levels of R&D and percentages of technical personnel on the work force were eligible for further preferential treatment.[102] In addition, foreign enterprises could participate in Torch Projects. By September 1996, 155 joint ventures with investment of $417 million were located in the zone.[103] As a balance to FDI, local leaders also concentrated funding from the Torch Program in the zone. By 1995, sixty-eight projects, or 68.7 percent of total Torch projects in Guangzhou, were in Tianhe. Science and Technology Commission officials argued that the reforms created an extremely "broad" and "diversified" market; to compete, the zone had to focus on three pillar industries—electronics and communications, biotechnology, and new materials. There has been a special focus on communications that make up 60 percent of all products.[104] Much of this focus, however, was extremely short term and the zone generally did not participate in programs where investment extended over two years or the investment-profit ratio fell below 1:5. As mentioned above, in 2000 zone management also began focusing on software development.

The zone has also expressed an interest in using Torch funds to develop technologies for traditional industries. This appears to be a large part of the zone's development strategy as demonstrated by the only 17 percent of enterprises in the zone in 1995 that were engaged in high-technology production.[105] Zone officials spoke of a "permeate and peel off" (*shen tou, bo li*) strategy.[106] The zone used preferential policies to allow new technologies and capital to "permeate" SOEs as well as to demonstrate the applications and markets for new products. The "peel off" strategy attempted to sepa-

102. "General Construction Situation," in *Investment Guide*.
103. "Preferential Policies for Foreign Joint Ventures," in *Investment Guide*.
104. Management Office of the Guangzhou Tianhe New High-Technology Industries Development Zone, "Zhua Huaju Xiangmu Longtou, Chuangjian Tese Gaoxinqu" [Seize the Torch Plan, establish a special new and high-technology zone], in *Shi Zhang Zuo Tanhui: Wenjian Zailiao Huibian*, 173.
105. "General Construction Situation," *Investment Guide*.
106. Ibid.

rate personnel and capital from SOEs and entice them to the zone to set up their own enterprises.

Finally, local officials used Torch Projects to support *minying* enterprises. Not only were these projects expected to funnel much-needed capital to nongovernmental enterprises, it was also hoped that participation in the project would build a modern enterprise system. Overall, zone officials adopted a management style that did not interfere with daily management decisions of enterprises; personnel decisions, profit distribution, and relations with supervisory agencies were all designated areas where zone officials could not interfere. Officials also used funds to set up model enterprises that demonstrated modern management and operating techniques in the hope of influencing and educating entrepreneurs in the zone.

Government Supervision in Xi'an

In Xi'an, as in Guangzhou, municipal leaders declared high-technology as one of the city's "pillar industries" and created high-technology zones to provide institutional support for technology enterprises. Both city and provincial governments officially managed the Xi'an High-Technology Zone, which was established in 1991, but the city was the dominant partner.[107] The two levels of government created the High-Technology Zone Small Leading Group, staffed by provincial and city officials responsible for S&T development and chaired by Xi'an party secretary Cui Lincou. The leading group was responsible for setting macroeconomic policy, deciding the size and scale of projects in the zone, and coordinating action between different parts of the S&T system. Below the leading group, the management office handled day-to-day management issues, project approval, real estate development, and residence permits. As in Beijing, the management office was an agency of the city government (*paichu jigou*) and possessed the city's highest economic authority. As one management office official described it, "What the city government can do, we can also do."[108]

Formal supervisory power was more concentrated than in Guangzhou, and the power was used, according to zone officials, "to break through the restrictions of the old system, including the barriers between different segments of the S&T system." But as the zone was laid out to increase the local government's ability to provide guidance, it could also lead to the over involvement of the local government in the management of the zone or enterprises. Officially, at least, local leaders resisted this temptation, describing their management strategy as "small government and large society,

107. In Chinese, "*sheng shi gong ying, shi ying wei zhu.*" See Management Office of the Xi'an New and High-Technology Development Zone, "Kua Shiji Chanye Jidi—Xi'an Gao Xin Jishu Chanye Kaifaqu" [The base of next century's industry—Xi'an new and high-technology industry development zone], in *Shi Zhang Zuo Tanhui: Wenjian Zailiao Huibian*, 224.

108. Interview, X2, 22 July 1998.

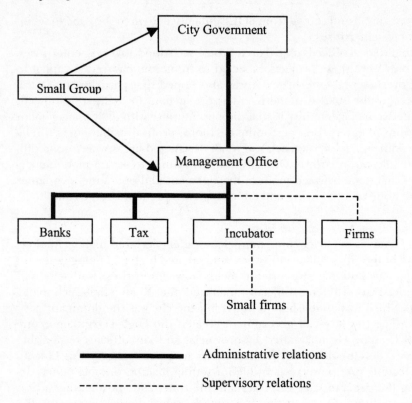

Figure 5.2. Xi'an High-Technology Development Zone

small organizations and large service."[109] Moreover, they described their management style as one that promoted the principles of nongovernmental organization, as defined by no outside interference and management autonomy, among as many enterprises as possible.[110]

Given their importance to the Xi'an economy, SOEs were a major target of local officials, who tried to use the zone to increase the rate and efficiency of the absorption by SOEs of new technologies. The zone promoted a "two ends in the zone, and the middle outside" strategy for SOEs, encouraging them to locate R&D and final parts processing in the zone and leave their factories and manufacturing outside of it.[111] More surprising was that local officials also argued that public-sector enterprises should relocate to do more than take advantage of the zone's preferential policies. They

109. Xi'an Management Office, "Kua Shiji Chanye Jidi" [The base of next century's industry], 222.

110. Interview, X2, 22 July 1998.

111. "Liang tou zai nei, zhong jian zai wai." Management Office of the Xi'an New and High-Technology Development Zone, "Jianshe Gao Xin Yuanqu, Zhenguang Shaanxi Sheng" [Build a new and high-technology park], in Quanquo Kexue Jishu Dahui [Documents from the national meeting], 155.

should also study the example of the zone's "model enterprises" and convert their management systems.

In 1993, Xi'an created a small business incubator, the Xi'an International Business Incubator, to funnel resources and provide institutional support to smaller nongovernmental enterprises. By 1997 there were 150 enterprises in the project generating an income of RMB 300 million.[112] The management office of the incubator involved itself in a range of projects, including creating an innovation fund, providing risk insurance, and helping *minying* entrepreneurs apply for city, provincial, and national S&T plans.[113]

Though officials in the incubator described their most important role as "creating a new way of thinking," the case of Xi'an Future International Software demonstrates the usefulness of the more prosaic services that the office provided. Future, a joint venture founded with a RMB 60 million investment from Xi'an Software Engineers and the Lion Group of Malaysia, created computer networks for banking and insurance companies.[114] Two students from the computer department of Xi'an Jiaotong University founded Xi'an Software Engineers in 1993; the two raised the start-up capital of RMB 200,000 from their savings, friends, and a small, nongovernmental investment company.[115] The company developed network software that was eventually used by the China Agricultural Bank in Hunan, Jiangsu, Hebei, and Gansu provinces. Growth was not smooth; some bank branches could not believe that software should be as expensive as it was and refused to pay.[116] But by 1995 the company had RMB 2 million in assets.

In explaining what role the incubator played in Future's success, managers listed three activities. First, and most important, the zone made it possible to get loans. Second, officials educated the public about what a nongovernmental business was and raised the social status of nongovernmental

---

112. Xi'an International Business Incubator, *Zhongguo Guoji Chuangye Fuhuaqi: Wunian Huigu* [China international business incubator: A five-year review] (Xi'an: n.p., 1998).

113. Jing Junhai and Jin Hui, *Keji Qiye Chengzhang yu Qiye Fuhuaqi* [The growth of science and technology enterprises and business incubators] (Xi'an: Xibei Gongye Daxue Chubanshe, 1998), 190.

114. "Software Venture Opens in Xi'an," *China Daily*, 13 December 1996. It is also interesting to note that Gold Lion, which was looking for an investment opportunity in IT, did not form a partnership with Haixing, supposedly Xi'an's most advanced technology enterprise. People at Future explained it by saying that they actually focused on IT; Haixing was too diversified.

115. Interview, X11, 29 July 1998.

116. This case provides a little more insight into the question of the role of personal connections, or *guanxi*, in the development of nongovernmental enterprises discussed earlier in chapter 3.

When I asked a manager at Future about how *guanxi* hurt or helped nongovernmental enterprises, he recounted a story of a branch of the Agricultural Bank that refused to pay more than RMB 200 for network software. By using personal connections, Software Engineers was able to get the central office of the bank to put pressure on the branch office. The company was eventually paid the full fee. But the manager also stated that the enterprise had to rely on these connections less and less often. Defining property rights more clearly had made it more likely that the enterprise could go through official channels.

entrepreneurs in society as a whole. Finally the zone maintained files on social insurance and other personnel matters. Before the zone took over
these responsibilities it was unclear to S&T personnel about what would
happen to their personnel files (*dangan*) if they left a state-owned research
institute for a *minying* enterprise.

Informal Supervision: Guangzhou

Guangzhou's most difficult problem was mobilizing human capital and
encouraging talented scientists and researchers to establish their own companies. The city did not start the reform process from a position of
strength. In a 1985 survey, Guangzhou ranked ninth on a list of sixteen
other large cities in terms of S&T personnel strength; S&T personnel made
up less than 3 percent of the total industrial work force.[117]

The city did try to encourage S&T personnel to moonlight and consult
for other enterprises. But these measures had a limited effect for three reasons. First, they had difficulty creating a social environment supportive of
"moonlighting"; scientists reported that their units looked down on the
practice, and individuals were unsure what would happen to their personnel dossiers if they left their old units. These scientists feared becoming "va-
grants" (*youmin*).[118] Second, it seems quite possible that a number of S&T
personnel were unaware of the policies. From 1979 to 1994, the local STC
issued seventy new laws and regulations on science and technology reform.
But in one survey, 30 percent of S&T personnel reported that they had no
idea what most of these policies were about and 40 percent responded that
they did not know how to find more information about new regulations.[119]

Finally, and probably most important, Guangzhou officials had to compete with Shenzhen and Zhuhai for talent. The preferential policies and
aggressive high-technology development strategies adopted by these cities
created a flow of skilled personnel—"peacocks flying southeast" (*kongque
dong nan fei*)—that affected Beijing, Xi'an, Guangzhou, and Shanghai. By
the time Guangzhou tried to retain S&T talent through measures like making it easier to obtain a city residence permit or an exit visa to Hong Kong,
many of the most entrepreneurial scientists had already left.

The problem of mobilizing S&T talent was closely related to the way the
Guangzhou officials politically and socially legitimated *minying* enterprises.
During the early 1980s, *minying* entrepreneurs registered as collective no
matter what their ownership structure; they, like entrepreneurs every-
where, "wore the red hat" (*dai hong maozi*) since private property was too

117. "Guangzhou Yunxu Keji Renyuan Ziji Xuanze Zhiye" [Guangzhou allows S&T personnel to choose their own careers], *Keji Ribao*, 16 July 1987.
118. "Fanghuo Keji Renyuan Kunnan You 'San' " [Three difficulties in enlivening S&T per-
sonnel], *Keji Ribao*, 29 May 1987.
119. Zhang Lifa and Xu Huxiang, "Guangdong Sheng Keji Renyuan Zhengce Zhixing
Zhong de Wenti" [Problems in implementing Guangdong's S&T personnel policy], *Keji Guanli
Yanjiu* 2 (1994): 47–50.

politically sensitive. Eventually, the local government moved aggressively to put an end to this practice. The status of private enterprises in all sectors was raised, and entrepreneurs were told to register their enterprises so they reflected their true ownership. In the long term, this is an important strength for *minying* enterprises in Guangzhou. As one local STC official argued, *minying* is a temporary concept, and clear property rights are fundamental to future growth. In the short term, however, this meant that the local government made little effort to distinguish nongovernmental technology enterprises from other nonpublic sector enterprises. Local officials were less likely to play the role of the "good mother-in-law," the official who helped entrepreneurs clear bureaucratic hurdles. As a result, social incentives as well as material ones encouraged talented S&T personnel to consult for JVs or leave for Shenzhen, not set up their own companies in Guangzhou.

Informal Supervision: Xi'an
Informal supervisory activities in Xi'an concentrated on encouraging horizontal links between S&T institutes in different administrative systems, as well as fostering the rational flow of personnel between these institutes and into new enterprises. One of the central explanations for Xi'an's inability to convert its strengths in scientific resources into economic growth was the separation of research institutes and personnel into two separate systems—one dominated by the Education Commission, the other the defense industries. The local STC, at either the provincial or municipal level, did not have the administrative power to coordinate action between the two bureaucratic systems.

In 1988 the Shaanxi Science and Technology Commission held a meeting about strengthening the authority of the science commission and created organizations intended to foster horizontal ties between military and university research institutes and production units.[120] These efforts did not progress far as demonstrated by the 1995 announcement of the creation of a Provincial Small Leadership Group for S&T. Chaired by the governor, with the heads of the Shaanxi Science and Technology and Economic Commissions as vice chairs, the group's mission was not only to increase funding and investment for technological development but also to coordinate and encourage central research institutes in the education and defense systems to participate in local development plans.[121]

Xi'an's other major concern was holding on to its skilled S&T personnel. In 1988, Shaanxi initiated a dual strategy of relaxing control on individuals in research institutes and encouraging them to set up their own en-

120. "Shaanxi Sheng Kewei Zuzhi Gelu Keji, Dajun Kaixiang Jingji Jianshe Zhu Zhanchang" [Shaanxi STC promotes science and technology, pushes forward the main battlefield of economic construction], *Keji Ribao*, 7 January 1988.

121. Shaanxi Provincial Government, "Fazhan Minying Qiye," in *Quanquo Kexue Jishu Dahui* [Documents from the national meeting], 152.

terprises. The local government also allowed enterprises to experiment with technology stocks earlier than in Guangzhou, and by 1995 Xi'an Jiaotong University had set up more than thirty enterprises that employed more than one thousand employees. The municipal and provincial governments aggressively recruited former Xi'an residents who had either gone to the coast or abroad to return to the city or at least to open branch offices in the high-technology zone.

The local government's role in political and social issues was murky during the early stages of reform. During the early 1980s, the local government worried about SOEs and did little to change the atmosphere of discrimination that predominated against *minying* enterprises. Entrepreneurs registered as collective to avoid any appearance of being private, and as one local entrepreneur stated, the local government was very clear about what type of enterprises it was going to support; "Nongovernmental was definitely 'white,' anything in the public or collective sector definitely 'red.' "[122]

Attitudes changed slowly. After the Tiananmen massacre, municipal officials described *minying* enterprises as a natural and desired product of reforms and argued that the "fate of the party and *minying* enterprises are tightly linked."[123] Later that summer, leaders further assured entrepreneurs that the policies on S&T reforms were not going to change.[124] In April 1991, during a larger meeting on S&T issues, a small meeting was held where entrepreneurs expressed concerns with the current development strategy to provincial leaders.[125] And in the course of 1992 and 1993, the municipal and provincial government passed laws defining, regulating, and supporting *minying* enterprises. In effect, Xi'an gradually changed the pattern of its support for *minying* enterprises. The city's promotion of Haixing was an important part of this process. Despite legitimate doubts about the indigenous technological capabilities of the company, Xi'an's celebration of Haixing at least sent the message to would-be entrepreneurs that *minying* enterprises could succeed locally.[126]

CONCLUSION

How local entrepreneurs interpreted this message is open to question. Other Xi'an entrepreneurs were quick to note the close ties between Haixing and the local government. In fact, the dominant message may be summed up by one entrepreneur's comparison of how *minying* enterprises

---

122. Interview, X6, 24 July 1998.
123. *Zhongguo Minban Keji Shiyejia Xiehui Huixun* [Newsletter of the China nongovernmental science and technology entrepreneurs association], 9 July 1989.
124. Ibid., 4 October 1989.
125. Ibid., 24 June 1991.
126. And I do mean celebrate. During the summer of 1998, city streets were filled with banners commemorating the ten-year anniversary of the enterprise.

grew in and outside of Beijing; "In Beijing enterprises were supported by the government, and then they grew. Here we grow, the government notices us, and then it supports us."[127]

From 1980 until 1993, Guangzhou and Xi'an adopted a "grow then support" attitude toward *minying* enterprises. These attitudes fostered two patterns of technological development in each city; Guangzhou has a more "market-oriented" pattern, Xi'an more science based. In Guangzhou, the local government focused on attracting FDI and fostering TVE growth. Technological development was assumed to be a natural and rather easy part of these processes. As a result, funding, property rights, and regulatory policies did little to identify and support the industry-specific needs of technology enterprises. Nongovernmental enterprises remained small and privately owned, often competing unsuccessfully with higher-return, lower-risk sectors for capital. Moreover, the slow pace with which local government officials went about defining and regulating nongovernmental enterprises severely handicapped the rate at which local entrepreneurs could develop more modern patterns of enterprise management. Unable or unwilling to create internal stock systems, individual scientists continued to manage more mature enterprises like the small collection of friends most *minying* enterprises were when they were founded. Many talented scientists left Guangzhou knowing that nongovernmental enterprises were better developed and better supported in Shenzhen and Zhuhai.

In Xi'an, city and provincial leaders tried to mobilize extensive local S&T resources. But this more science-based strategy was hampered by the need to maintain the position of state-owned enterprises in the local economy; local planners approached *minying* enterprises as a complement to the public sector. As in most other areas of reform, Shaanxi provincial and Xi'an municipal leaders proceeded extremely slowly and cautiously with supporting nongovernmental enterprises. Funding was focused on the state sector, and planners had difficulty coordinating a large, but disjointed S&T system. Limited support meant that entrepreneurs had difficulty securing access to capital. Those that stayed in Xi'an adopted strategies based on exploiting one market niche, quickly diversifying, and fostering close ties with local government authorities.

Policies in both areas were to change as local planners in Guangzhou grew skeptical about the ability of FDI alone to raise indigenous technological capabilities, and officials in Xi'an became frustrated with the complexities of reforming moribund SOEs. The future of development in both areas is uncertain; Xi'an, even with a significant S&T base and a policy environment supportive of nongovernmental enterprises, may continue to suffer from the relative absence of capital. Skilled S&T personnel may continue to leave Shaanxi for more developed areas. The "go west" development plan does not appear likely to attract significant FDI. Guangzhou may

127. Interview, X4, 23 July 1998.

not be able to overcome its lack of S&T resources. The city may in the future not only trail Beijing and Shanghai but also other cities in Guangdong.

What both cases do show is how much more open the range of policy choices have become in China. The development experiences of different nations, in addition to different regions within China, have slowly diffused throughout the country and been interpreted and implemented to varying degrees. The central government created the large environment for reforms but within that space local officials had room to maneuver. Moreover, many of them looked to California or Hsinchu Science Park in Taiwan or the Beijing Experimental Zone for ideas about how to organize innovation. Technological development in Guangzhou and Xi'an is not simply a story of how *minying* enterprises were regulated; it is also a story of how the development experiences of different areas were diffused, adopted, and implemented.

CHAPTER SIX

Technology and China's Future

At the end of 1996, the world of nongovernmental enterprises received a shock: Giant (*Juren*), one of the most famous and successful nongovernmental enterprises, essentially went bankrupt. Founded in 1989 by Shi Yuzhu with an original investment of RMB 4,000, Giant had sales of its Chinese language card totaling RMB 160 million and profits exceeding RMB 35 million by 1992.[1] The company set up an office on Beijing's "Electronics Avenue" and hired one of the former managers of Founders, Luo Binlong, to head the new Giant Advanced Technology Group. Both Jiang Zemin and Li Peng visited the company and praised it as a model for other nongovernmental enterprises throughout the country.

By 1994, the company's fortunes had begun to fade. Facing increased competition, Giant made a number of risky investments in high-tech products like hand-held computers did not pay off. Talented S&T personnel left the enterprise, and Microsoft sued the company for intellectual property rights violations. Giant also diverted increasing amounts of money into real estate, using RMB 1.2 billion to build what at the time was to be China's tallest building, investing another 480 million in an amusement park, and purchasing 30,000 square meters of land in Pudong. In 1995 the company attempted to revive itself with three new products (one computer related, the other two health food goods), but these products had little market share, and the large advertising campaigns that accompanied their launch increased the enterprise's debt burden. An increase in the number of employees from one thousand to three thousand in a single month created internal management chaos. By January 1997 Giant owed billions of renminbi to investors on the mainland and in Hong Kong, and officials in Zhuhai made it clear that they had no responsibility to bail out the foundering company.[2]

Analysts on the mainland listed numerous and sometimes contradictory reasons for Giant's failure.[3] The company never grew large enough, or it grew too large, too quickly. Management authority remained concentrated in the

---

1. This account of Giant draws heavily from "Ruhe Kandai 'Juren' Jianshi" [How to look at the "Giant" incident] *Beijing Keji Ribao* [Beijing science and technology daily], 14 July 1997.

2. Wendy Lim Wan-Yee, "Debt-Stricken Firm Holds Crisis Talks," *South China Morning Post,* 23 January 1997; and Ada Yuen, "Giant Boss Admits to 'Major Mistake,'" *South China Morning Post,* 3 February 1997.

3. "Ruhe Kandai 'Juren' Jianshi" [The "Giant" incident].

hands of Shi Yuzhu. Despite Jiang Zemin and Li Peng's visit, nongovernmental enterprises did not have the status of state-owned enterprises and could not secure bank loans. Perhaps most important, property rights were never clearly defined within Giant, and the enterprise was not allowed to issue shares and list on the stock market. Shi Yuzhu himself argued that Giant's inability to raise capital with equity had led to an overreliance on debt.

Whatever the reasons for failure, Giant's story was familiar to high-technology entrepreneurs in Beijing, Shanghai, Xi'an, and Guangzhou. Giant's rapid growth from a small enterprise in Zhuhai to national prominence was a microcosm of the difficulties of creating a high-technology enterprise in China. The rate of Shi Yuzhu's rise and the heights from which he fell may not have been typical for most entrepreneurs, but the problems he faced in creating and managing his enterprise were. Giant's path was not unique; enterprises throughout the country adopted many of the same strategies.

The achievements of those entrepreneurs that do succeed are even more remarkable given the barriers they confront. Individuals like Liu Chuanzhi of Legend or Deng Longlong of Suntek founded enterprises in an environment lacking most of the economic, political, and social institutions that support a modern firm system. The first successful nongovernmental entrepreneurs built enterprises responsive to rapid market changes without the assistance of independent industrial associations, a well-developed legal system, or clear property rights.

THE ROLE OF LOCAL GOVERNMENTS

The absence of civic organizations and other social institutions meant that government actors played a more observable and central role in the construction of new technology sectors in China than they have in the West. Government actors, at the national and state level, are an often-overlooked part of the story of Silicon Valley and Route 128. But even if more attention is paid to the role of California officials, there is no equivalent in Palo Alto or San Jose to the official who acts like a "good mother-in-law" in Haidian district. Science and Technology Commission officials in Beijing fulfilled a whole range of roles, such as introducing prospective business partners and organizing job fairs for recent graduates, considered the responsibility of private actors in the West.

The political roles and economic strategies government actors adopted shaped the emerging information industry sector within specific localities. Local government officials rarely asked themselves if they should or should not intervene in new markets. Rather, the question was whether they would choose to intervene in ways that supported technological innovation: would government actors help entrepreneurs secure access to scarce inputs while maintaining enterprise autonomy? In China, local officials made decisions within three policy arenas: investment structures, property rights, and government supervision. How (and when) local authorities chose to define nongovernmental enterprises, what types of technologies they funded, and how often they interfered with internal business operations all affected the availability of critical inputs, the ability of managers to decide business strategies, and thus the competitiveness of local enterprises.

Funding was the most direct way to shape enterprise behavior. In Beijing, the local government provided more loans to smaller enterprises; Torch funds went to the most promising nongovernmental enterprises and funding continued until these enterprises could bring new products to market. By contrast, the Shanghai local government directed foreign investment to a few large state-owned enterprises. As a result, foreign investment was much larger in Shanghai than in Beijing. Guangzhou and Xi'an each adopted different investment strategies. With easy access to Hong Kong, Guangzhou officials directed foreign direct investment into smaller collective enterprises. These investments provided needed operating capital for the local partner, but did little to raise indigenous technological capabilities. Xi'an officials were not as fortunate in their access to foreign funds and so tried to take advantage of the funding already present in local public-sector institutes.

Local government decisions about which type of property rights to acknowledge also directly influenced enterprise structure and thus enterprise competitiveness. In Beijing, local officials used the 1984 "Inquiry into Reforming Beijing's S&T System" to recognize the role nongovernmental enterprises were playing in technological development. In general, local officials recognized multiple forms of ownership and encouraged collective enterprises to adopt stock systems. The priority for Shanghai officials was to use stock systems to rejuvenate and reorganize state-owned enterprises; nongovernmental enterprises remained outside of their concern. Local officials in Xi'an adopted a similar attitude toward property rights, using limited stock companies to promote SOE reform. And in contrast to the rest of China, officials in Guangzhou were the most aggressive about defining as many nongovernmental enterprises as private as possible.

Finally, local governments had to decide how to supervise nongovernmental enterprises. Activist governments, like Shanghai and Xi'an, maintained an ownership share and a position on the board of directors in some companies. These governments tended to see nongovernmental development as a complement to the larger public-sector economy. Most supervision fostered vertical ties between state actors. By contrast, the Beijing government placed nongovernmental enterprises at the center of local development strategies, fostering horizontal links among a whole range of state and societal actors in an attempt to create an "innovation network" (*chuangxin wangluo*). The local government tried not to interfere in internal business administration and protected enterprises from outside interference from other administrative agencies. Again, the Guangzhou local government took the most laissez-faire stance on nongovernmental enterprises, appearing unconcerned with either fostering links between actors or supervising enterprise growth.

These government actions resulted in distinct patterns of nongovernmental enterprise development. These patterns differed in terms of enterprise size and ownership structure, the quantity and quality of local government ties, and the extent to which government action fostered horizontal ties between actors. More flexible forms of organization supported by interpersonal links characterized Beijing. Enterprises in Beijing were linked to a wide range of supervisory agencies; enterprises spun off from the Chinese Academy of Sciences (CAS) developed additional ties to local SOEs and independent research institutes, diluting the administrative control CAS had over business strategy. These links created positive externalities, provided public goods, and facilitated information flows, and thus were more likely to support innovation. School networks linked enterprises to each other and to local and central government officials, and entrepreneurs began building their own industrial associations and consulting enterprises like Taisun and Great Wall.

Shanghai was dominated by a smaller number of large corporations that fragmented technological networks. Small private and collective enterprises were separated from each other and from government actors and so less likely to develop the relationships that raise the competitiveness of any one enterprise operating alone. Large SOEs had ties to the local government that not only reduced the flexibility of individual enterprises, but also allowed SOE managers to lobby for continued preferential treatment. Enterprises often had more than one representative of local government on the board, and these officials frequently influenced internal business management decisions. Unlike in Beijing, the local government did not make the protection of nongovernmental enterprise autonomy a priority.

The municipalities of Guangzhou and Xi'an possessed some of the factors that support external economies but lacked others. Enterprises in Guangzhou tended to be small, relatively isolated, and privately owned. Connections, cooperative or competitive, were better developed with firms from Hong Kong and Taiwan than with other local enterprises. Most Guangzhou enterprises started with few contacts with the local government, though they began developing more extensive ties in the late 1990s. In Xi'an, the development of networks between nongovernmental enterprises was hampered not by the presence of joint ventures, but by the large public sector and defense industry infrastructure. Small privately and collectively owned enterprises have historically been separated from each other and from important government actors. But, like in Guangzhou, this began to change at the end of the 1990s. Local officials supported enterprise development in return for authority over internal business management.

LOCALITIES AND INSTITUTIONAL CONSTRAINTS

Decisions about how to implement property rights reform, technology funding, and government supervision directly affected the organization of nongovernmental enterprises and regional competitiveness in the information technologies. So why did localities adopt the strategies that they did? More specifically, why was Beijing more successful in combining guidance and support for entrepreneurs while still allowing individual effort to shape the sector? In each locality, organizational resources constrained policy choices. In the four cases discussed in this book, the local economy's role in the national economy, the presence or absence of S&T resources, FDI, and skilled personnel, and the degree of autonomy and cohesion of the local government were all crucial factors in determining which policy instruments were available.

Why did the municipal bureaucracy in Shanghai act in a unified manner directing foreign investment and other resources to the state-owned sector? Shanghai's organizational resources heavily favored a development approach that emphasized government planning. Shanghai, more than any other local economy, was a planned economy, and the local economy was extremely important to the national planned economy. As Lynn White argues, in Shanghai the central planning system was at its most extensive, and factory managers were limited not only by more rules and regulations but also by more attentive state agencies.[4] Neither central planners nor local officials in Shanghai had incentives to support nongovernmental entrepreneurs, and the city lacked a social base of technological entrepreneurship.

Beijing officials could never duplicate the degree of control their counterparts in Shanghai had over the local economy. Even if they had desired to coordinate activity between technology enterprises, Beijing local officials lacked the necessary policy tools. Nongovernmental enterprises existed in a space between the central and local governments, and technological entrepreneurs had tight connections to officials at both levels. These personal ties were important for more than economic reasons; they also gave entrepreneurs in the capital a better sense of which direction the political winds were blowing and more confidence that the central government intended to follow through on policies supportive of nongovernmental enterprises.

Both its distance from central control and its proximity to Hong Kong heavily influenced Guangzhou's development path. During the 1980s Guangzhou remitted a higher percentage of its revenue than any other city in Guangdong to the provincial government. What money Guangzhou did

---

4. Lynn T. White, *Shanghai Shanghaied? Uneven Taxes in Reform China* (Hong Kong: University of Hong Kong, 1989), and idem, *Unstately Power,* vol. 1, *Local Causes of China's Economic Reforms* (Armonk, N.Y.: M.E. Sharpe, 1998).

control, it funneled to large infrastructure projects or real estate development, not the S&T budget. Nongovernmental entrepreneurs had to compete with firms from Hong Kong, Japan, and Taiwan, and the neighboring
cities of Shenzhen and Zhuhai lured S&T talent away by offering attractive
tax breaks and other subsidies for technology enterprises. The position of
nongovernmental enterprises in Xi'an cannot be divorced from the lack of
foreign investment, the dominant position of SOEs, and the role defense
industries played in the local economy. Local officials were torn between
wanting more scientists to start their own companies and the need to protect the technological capabilities of state-owned enterprises.

LOCAL CULTURES AND DEVELOPMENTAL STRATEGY

The Chinese example of gradual reform, of attacking some problems first
while leaving more daunting challenges for later, has often been contrasted with the one-shot approach to reform adopted in some countries in
Eastern Europe. The Chinese removed price controls on some products
relatively quickly, but the reforms of SOEs have been much more tentative.
By contrast, policymakers in Russia generally tried to reform the economy
as quickly as possible across as many sectors as possible. Macroeconomic
stabilization, trade liberalization, and privatization of state-owned enterprises were all to be completed simultaneously.

Even if China had adopted a less incremental, more "big bang" approach to transition, there are good reasons to believe that organizational
constraints would have still influenced policy choices. Reducing the government's role in the economy does not mean eliminating the influence of
the economic and political institutions of the previous system. As David
Stark describes in Eastern Europe, "Actors in the postsocialist context are
rebuilding institutions and organizations not *on* the ruins but *with* the ruins
of socialism as they redeploy available resources in response to their imme-
diate practical dilemma."[5] The Chinese case reinforces this argument. The
choices that local governments made in regard to nongovernmental enterprises were highly path dependent. Local officials built new IT sectors with
the institutional resources distributed to them by the central plan. Without
its large concentration of universities and public research institutes, Beijing
would not have been able to support a vibrant nongovernmental sector.
Local officials in Xi'an and Shanghai had little choice but to try and link
emerging nongovernmental enterprises to research institutes connected to
the defense industries or located within SOEs, and it made sense for
Guangzhou to look over the border to Hong Kong.

The case of IT in China suggests that path-dependent approaches must

5. David Stark, "Recombinant Property in East European Capitalism," *American Journal of Sociology* 101, 4 (January 1996): 995.

encompass both the visible institutions in each locality as well as the ideas that animated these institutions. Simply looking at organizational structures only provides the broadest outline of the possible development paths local governments could have followed. Tracing the specific policies created in each locality requires placing each of the four cases discussed in this book within its distinct local culture. These cultures consisted of shared meanings and institutionalized practices that addressed the role of local governments, the organization of enterprises, and the use of horizontal and vertical coordination in new markets. These understandings helped identify the problems local officials wanted to solve and suggested culturally conceivable solutions to those challenges. Local officials in Shanghai, for example, were more likely to see creating large enterprise groups as a solution to coordination problems in the IT industry than their counterparts in Beijing or Guangzhou. The point is not that these local cultures existed independently of material constraints. Rather they were woven into the fabric of local institutions and gave meaning to organizational arrangements, linking desired economic outcomes to specific policy tools.

EXTENDING THE ARGUMENT TO OTHER SECTORS: COMPARING AUTOS AND IT IN TWO CITIES

In his study of railway policies in Britain, France, and the United States, Frank Dobbin argues, "History has produced distinct ideas about order and rationality in different nations, and modern industrial policies are organized around those ideas."[6] History has produced distinct ideas of how to order economies at the local level as well. Comparing how these ideas reveal themselves in industrial sectors as different from each other as high technology and automobiles suggests that Dobbin's argument can be extended; what Dobbin shows to be true across countries is also true across regions within them. The central government created institutions that reflect Chinese conceptions of order and rationality but gradually over the history of People's Republic of China these ideas were modified by economic and political actors and shaped to fit the needs of localities, creating distinct regional economic cultures. Confronted with new development challenges like developing an automobile or IT sector, local officials often relied on traditional practices of development to deal with uncertainty; industrial strategies were reproduced when localities tackled new problems

These strategies reveal themselves even in sectors that require different types of government support. In high technology, most successful firms are small, capital requirements low, and innovation rapid. High-technology firms benefit from horizontal and often informally organized interpersonal

---

6. Frank Dobbin, *Forging Industrial Policy: The United States, Britain, and France in the Railway Age* (New York: Cambridge University Press, 1994), 2.

linkages. Within firms, authority is often widely dispersed throughout the organization. By contrast, in autos, firms tend to be extremely large, start-up capital requirements huge, and the industry characterized by incremental learning rather than rapid innovation. The auto industry requires a vertical organizational structure that monitors production standards across individual firms tightly linked by supply networks. Small firms often do not have the technical skills, managerial ability, or market information necessary to increase their manufacturing capabilities, and it is extremely important that a group of firms develops together. In other words, computer software and automobiles clearly demand different strategies regarding organization, scale, and capital.

The Beijing and Shanghai local governments approached building a car and developing internet software as if they were basically the same problem.[7] Each local government reproduced distinct patterns of behavior in both sectors although these routines were not equally appropriate for the new industries. Shanghai officials were likely to see large conglomerates and government coordination as the way to organize new markets. Local leaders were accustomed to intervening in the market; the managers of state-owned enterprises were used to listening. In contrast, Beijing municipal officials relied on smaller enterprises and a more hands-off approach to market coordination for both autos and information industries.

As Eric Thun describes, in the early stages of developing the auto industry the Shanghai local government played a pivotal role with respect to capital accumulation and investment.[8] In 1986, the local government formed an Automobile Industry Leading Small Group directly under the auspices of the mayor's office to coordinate and monitor development efforts. Below this group, the local government also created a localization office to deal with day-to-day issues. In 1988 the Shanghai government began the process of capital accumulation by creating a "localization tax" on each Santana automobile sold. Proceeds made up a localization fund controlled by the municipal government and then reinvested in Shanghai producers. In addition, the local government organized a hierarchical conglomerate, the Shanghai Automotive Industry Corporation (SAIC), which facilitated learning, coordination, and development among supply enterprises. SAIC monitored the performance of and determined how much profits could be retained by individual enterprises; it also appointed the managers. Moreover, SAIC served as the planning base for enterprises within the group, complementing the role of the localization office. The bureaucratic structures of SAIC and the local government insured that top municipal leaders were both informed about and involved in development decisions.

7. This section draws heavily on Adam Segal and Eric Thun, "Thinking Globally, Acting Locally: Local Governments, Industrial Sectors, and Development in China," *Politics & Society* 29, 4 (December 2001): 557–88.

8. Eric Thun, "Changing Lanes in China: Reform and Development in a Transitional Economy" (Ph.D. diss., Harvard University, 1999).

In Beijing, the automobile sector did not become a "pillar industry" of the local economy. The municipal government failed to create organizational structures conducive to collaborative relationships among enterprises, or to raise the funds needed for industrial upgrading. Many of the state-owned supply enterprises within the Beijing conglomerate were not wholly owned by the city. Use of multiple suppliers and commercial relations between enterprises limited the ability of supply enterprises to develop technical and manufacturing skills. The Beijing Automotive Industrial Corporation (BAIC) made no effort to pool knowledge within the head office, and BAIC was unable to help suppliers overcome marketing, management, or technical difficulties. In addition, the city government took a relatively laissez-faire approach toward investment and auto industry development, refusing to enact policies that would favor Beijing manufacturers.

These distinct local policies, Thun argues, resulted in different outcomes for the automobile market. By 1997 the manufacturing capability of the Shanghai automobile industry had increased sharply. Shanghai manufacturers were supplying almost 83 percent of the parts used to produce the Santana, and Shanghai manufactured 52 percent of all sedans sold in China. In Beijing, local content rates have not increased over 30 percent, and rates of production of the Jeep Cherokee have not kept up with the Santana.

Shanghai has trailed behind Beijing in new technology sectors because it adopted an approach that varied little from the one used to create its automotive success. What worked for Shanghai for autos has failed so far for information industries, and what failed for Beijing in autos has worked for new technologies. Again, it is important to note that these development strategies emerged not from institutional arrangements or local development practices alone, but from the interaction of the two. Shanghai did not have an institutional framework supportive of small and autonomous enterprises, and the Shanghai government organized enterprises into large groups. Tackling new development challenges, bureaucrats tried to coordinate development across enterprises as a complement to broader efforts in the locality. In Beijing, political and institutional structures gave entrepreneurs a degree of autonomy that stimulated informal ties and innovation. Confronting new industries, local officials reproduced a relatively hands-off approach that led to a flourishing high-tech sector but hindered the development of the auto industry.

LOCAL GOVERNMENTS AND NATIONAL DEVELOPMENT

Comparing local development in Beijing, Shanghai, Guangzhou, and Xi'an, either in automobiles or information industries, reinforces many of the lessons drawn from larger comparative studies of Japan, Korea, and

Taiwan about the role of government intervention in development. The studies of the developmental state in East Asia by Chalmers Johnson, Alice Amsden, Thomas Gold, Robert Wade, and others have been central in defining and illustrating a positive role for government bureaucracies in development. More recent work on development have swung the pendulum back once again, stressing the importance of private actors and enterprise networks to successful state intervention.[9] There is, however, an emerging consensus, well illustrated by the experience of the information industries in China, that markets require supporting institutions.[10] Neither private enterprises nor government actors can go it alone. New enterprises need the discipline of the market, but markets need constructing. Without clearly defined property rights, transparent governance structures, and standardized management practices markets are unlikely to operate efficiently.

Previous work on East Asian development has tended to assume that the construction of these market-supporting institutions has occurred at the national level. This work has focused on highly visible components of the central government like the Ministry of International Trade and Industry in Japan or the Economic Planning Board in Korea. The state is reduced to administrative or policy organizations headed and coordinated by an executive authority.[11] IT development in Beijing, Shanghai, Xi'an, and Guangzhou suggests that this view of the state intervening in the national economy is overly monolithic. The Chinese state is not a clearly demarcated entity, separate from civil society and located in the office buildings of the central ministries on the streets of Beijing. Instead it is a set of relationships that encompasses central, provincial, and local government actors. As I have shown, an understanding of development in China rests on an understanding of local institutions, local constraints, and local politics.

This argument is not restricted to China. The process of development, particularly in large countries, is driven at both the national and subnational level. As AnnaLee Saxenian demonstrates in her study of Silicon Valley and Route 128, even in advanced economies like the United States and within the same sector there are important regional sources of competitive advantage.[12] The point is not simply that looking at regional economies provides a level of detail and nuance not available in studies focusing on

9. See chap. 1, nn. 4–7; Stephan Haggard, "Business, Politics, and Policy in Northeast and Southeast Asia," in *Business and Government in Industrializing Asia*, ed. Andrew MacIntyre (Sydney: Allen and Unwin, 1994); and Chung-in Moon and Rashemi Prasad, "Networks, Politics, and Institutions," in *Beyond the Developmental State: East Asia's Political Economies Reconsidered*, ed. Steve Chan, Cal Clark, and Danny Lam (New York: St. Martin's, 1998), 9–24.

10. World Bank, *The East Asian Miracle: Economic Growth and Public Policy* (New York: Oxford University Press, 1993).

11. This definition of the state comes from Theda Skocpol, *States and Social Revolutions: A Comparative Analysis of France, Russia, and China* (New York: Cambridge University Press, 1979), 29.

12. AnnaLee Saxenian, *Regional Advantage: Culture and Competition in Silicon Valley and Route 128* (Cambridge, Mass.: Harvard University Press, 1994).

national economies. Rather the lack of uniformity at the regional level in many economies makes local-level analysis a necessity. Institutions at the local level will often determine whether development policy succeeds or not. Moreover, as Saxenian further argues, regions may become even more important as production and markets become increasingly global. When production is embedded in regional social structures and institutions, enterprises can translate local knowledge and relationships into innovative products and services competitive in the world economy.[13]

For outside observers of China this complicates our understanding of development within the country, both political and economic. The reality may be that China is both farther ahead and farther behind where we would expect from looking solely at central government policy. On one hand, local governments may pursue more market-oriented polices than those officially supported by the center. This has happened throughout the reform period and complicates any one picture of how "free" or "open" the Chinese economy is at any one time. In fact, there may be significantly more space for private entrepreneurs at the local level than appears possible from the laws promulgated by Beijing.

On the other hand, the creation of national level policies in and of themselves may not be a particularly helpful marker of where the Chinese economy is. Supporters of China's accession to the World Trade Organization (WTO) and continued engagement with China more broadly have rightly argued that further enmeshment with the world economy will make China more transparent. Not only will the WTO force China to further liberalize its domestic industries, but China will also have to create national laws and procedures to regulate both foreign and domestic companies. As former secretary of the treasury Robert Rubin argued in congressional testimony about granting Permanent Normal Trade Relations to China, "By helping to open and liberalize China's economy, WTO accession will promote economic freedom and the rule of law in many sectors of the Chinese economy that are now dominated by state power and control. Compliance with WTO provisions will require reform in many areas of the Chinese economy and will require China to implement new laws and procedures that comply with WTO rules."[14] This is likely to be true, but to understand how these new laws and procedures actually work themselves out and affect development after China joins the WTO we will need to examine how these national policies intersect with local development.

TECHNOLOGY AND TECHNONATIONALISM

Although political scientists have paid great attention to the role of the state in the East Asian miracles, they have been less concerned about why

13. Ibid., 161.
14. Robert Rubin, Testimony before the House Committee on Ways and Means Regarding China's Accession into the World Trade Organization, 3 May 2000.

the state intervened in new markets. For the most part, promoting economic development and raising standards of living has been seen as reason enough. But, especially in East Asia, economic issues have often interacted with security concerns. Karl Fields, for example, argues that Korean and Taiwanese policymakers had to balance the needs of economic enterprises with strategic issues.[15] In the 1960s, Korean military officers and technocrats viewed rapid economic growth as essential to state security. Concerned about military threats from the North, South Korean leaders believed they needed a "big push" to speed heavy industrialization. Large business groups were considered the most effective way to achieve these goals, and technocrats offered massive subsidies and other incentives to the *chaebol*, a handful of large conglomerate groups. Owned and controlled by a single family, these groups were diversified across several industries and were closely linked to the government. The *chaebol* soon dominated the economy.

On Taiwan, the transplanted Nationalist regime adopted an industrial policy designed to maintain price stability and prevent the concentration of private financial power. Security concerns were both external and internal; in addition to the threat of the Chinese Communist party across the Taiwan Straits, the mainland Chinese Kuomintang was an ethnic minority regime on the island. The KMT manipulated financial institutions to prevent the emergence of large business groups tied to the indigenous majority and created an environment in which Taiwanese business groups were not as central to the domestic economy as *chaebol* were in Korea. Taiwanese firms pursued a strategy of cooperation with MNCs as a means to acquiring critical technologies, enhancing their own self-reliance from the Nationalist state, and offsetting Japanese influence over Taiwan's industrial structure.[16]

The link between economic development and state security is most clearly explored in Richard Samuels' work on "technonationalism" in Japan.[17] According to Samuels, Japanese technonationalism consists of three parts: indigenization, diffusion, and nurturance. Indigenization refers to the Japanese belief that the country had to achieve independent technological competence in order to achieve military power and security. Second, the Japanese established an extensive network of institutions that encouraged the diffusion of technology across industrial sectors. Finally, specific technologies were nurtured and producers and users of these technologies supported by the state. Japanese policies targeted not specific sectors, but capabilities; the knowledge generated by these industries was as important as the products themselves, if not more so.

15. Karl Fields, *Enterprises and the State in Korea and Taiwan* (Ithaca, N.Y.: Cornell University Press, 1995), 238.
16. Denis Fred Simon, "Taiwan's Technological Future: The Impact of Globalism and Regionalism," *China Quarterly* 148 (December 1996): 1199.
17. Richard Samuels, *Rich Nation, Strong Army: National Security and Technological Transformation in Japan* (Ithaca, N.Y.: Cornell University Press, 1994).

Like Japan, Taiwan, or Korea, China has also expressed security concerns through technology policy. Most prominently, Chinese leaders continue to worry about technological dependence on other countries, especially the United States. China has maintained a significant R&D effort in military-related sectors, and a focus on "critical technologies" has also spilled over into the civilian economy.[18] The importance of these types of research projects is sure to increase as a result of the U.S. bombing of the Chinese embassy in Belgrade, the release of the Cox Report accusing China of nuclear espionage, and continued tension across the Taiwan Straits. But, as this book has shown, China has gradually forsaken many of the institutions traditionally associated with technonationalism. Central planning has been scaled back and support for large state-owned enterprises diminished. Central leaders have embraced actors—like nongovernmental entrepreneurs—who were ideologically suspect before. What matters now for a national champion is not that it is state, collective, or privately owned, but that it is Chinese. In this context it is important to stress the strategic context of technonationalism rather than the policy tools through which these concerns are usually expressed.[19] Technonationalism as a policy orientation toward autonomy and independence from other states can be revealed though policies supportive of either state-owned or nongovernmental enterprises.[20]

This point may be best illustrated by the government's support of software companies like Red Flag and other nongovernmental enterprises working on Linux operating software. The promotion of the open Linux operating system has both economic and security motives. China is currently unable to compete in the domestic market for Windows-related software, and Chinese computer programmers and companies might play a much bigger role in the Linux world.[21] Promotion of Linux may also ease Chinese fears of dependence on Microsoft and the United States. Moreover, the Chinese government also suspects that Windows has "back doors" that allow the company or the agencies of the U.S. government to spy on users. In an editorial on "information colonialism" the *People's Liberation Army Daily* argued that China must develop its own software since "without

18. Evan Feigenbaum, "Who's Behind China's High-Technology 'Revolution'? How Bomb Makers Remade Beijing's Priorities, Policies, and Institutions," *International Security* 24, 1 (summer 1999): 95–126.

19. This point is further developed in Barry Naughton and Adam Segal, "Technology Development in the New Millennium: China in Search of a Workable Model," in *Crisis and Innovation: Asian Technology after the Millennium*, ed. William Keller and Richard Samuels (New York: Cambridge University Press, forthcoming).

20. All economic nationalisms specify a direction for foreign economic policy away from an "other" and lead governments to interpret their economic dependence on some states as a security threat. See Rawi Abdelal, *National Purpose in the World Economy: Post-Soviet States in Comparative Perspective* (Ithaca, N.Y.: Cornell University Press, 2001).

21. G. Pierre Goad and Lorien Holland, "China Joins the Linux Bandwagon," *Far Eastern Economic Review*, 24 February 2000; "IBM, Redflag-Linux Team on Linux Development," *China Online*, 24 August 2000.

information security, there is no national security in economics, politics, or military affairs."[22]

The companies that the central government believes will be able to compete with Microsoft and provide information security are no longer necessarily state-owned enterprises. Red Flag is backed by the Chinese Academy of Sciences. While the Chinese government's role in technology development is now more restricted, its motivation is perhaps more explicitly nationalistic than it has been in the past. In part, this is because of the current importance of nationalism to Chinese domestic politics. Now that communism has lost most of its ideological force in China, the party relies on nationalism to justify its continuing role in economic and technology policy.

CHINA RISING

Technological issues touch on more than questions of economic autonomy; technological concerns are also tightly linked to military issues. The success or failure of the attempt to create an indigenous technological capability will significantly impact China's ability to assert itself militarily in the world. Already the technological capability China does have makes it a more formidable power than most developing countries. Chinese writing after the Kosovo campaign noted that China was not like Iraq or Serbia; China's possession of nuclear and space technologies meant that Beijing had a far greater ability to disrupt U.S. military plans. Even before the Kosovo and Taiwan Straits crises, the government increased spending for military research and development under the premise that the People's Liberation Army needed an advanced scientific base to develop sophisticated weapons.

As I have shown, there are still significant barriers to technological development in China, and the presence of these obstacles must raise doubts as to whether the Chinese will be able to build the technological foundation for advanced weapon systems. In the West, China's future ability to project power is generally assumed.[23] In most of the "China threat" literature, China is able in the not too distant future to convert rapid economic growth into military and political power.[24] This new China may eventually threaten both regional and global order. Nicholas Kristof, for example, argues, "Almost nothing is so destabilizing as the arrival of a new industrial

22. "Guanzhu 'Xinxi Zhimin Zhuyi' Xianxiang" [Concerning information colonialism], *Jiefangjun Ribao*, 8 February 2000.

23. Except perhaps in Gerald Segal, "Does China Matter?" *Foreign Affairs* 78, 5 (September/October 1999): 24–36.

24. Thomas Christensen describes how China may complicate U.S. defense planning not by "catching up" but by developing technologies and strategies that focus on potential adversaries' weaknesses. See, "Posing Problems without Catching Up: China's Rise and Challenges for U.S. Security Policy," *International Security* 25, 4 (spring 2000): 5–40.

and military power on the international scene."[25] Rapid economic development has not only significantly raised the standard of living in China, but it has also allowed China to finance an extensive military build-up. While other countries in the region have reduced their military budgets, from 1988 to 1993 the Chinese spent $18 billion a year; in March 2001, the Chinese announced they would increase defense spending by 17.7 percent, the biggest expansion in real terms in twenty years.[26] Comparing China to nineteenth-century Germany and twentieth-century Japan, Kristof reminds us of "the difficulty that the world has had accommodating newly powerful nations."[27]

This is not the entire story. Rapid economic growth will certainly support some level of military modernization, and the Chinese have purchased a number of advanced weapon systems including the Su-27 and Su-30 fighters and the Sovremenny-class destroyer from Russia. But China will continue to remain two or three decades behind the United States and other Western powers unless it develops an advanced scientific research foundation. The political scientist Avery Goldstein notes that "despite impressively robust economic growth, there is little likelihood that Beijing can greatly accelerate this modernization process, mainly because China has not yet established the necessary world-class scientific research and development infrastructure."[28] Much of Japan's success in the international system stemmed from its ability to create a unified ideological vision of the relationship between technology and security and the institutions to support that idea. By contrast, the Chinese have been struggling with a coherent vision and the institutions to support it for at least 150 years.[29]

There is no guarantee that the current generation of China's leaders will be able to create the institutions that link technology to security in the next several decades. The context in which central policymakers struggle to create a coherent vision of a national technology policy has become increasingly complex. Past strategies of relying on large government-sponsored business groups have not succeeded on any significant scale, and current development plans include an increasingly broad-based view that the key actors in the next phase of technological development are likely to be nonstate entrepreneurs. Entrance into and compliance with WTO provisions that limit the ways the government can intervene in the economy will further complicate the decision-making process.

25. Nicholas D. Kristof, "The Rise of China," *Foreign Affairs* 72, 5 (November/December 1993): 60.

26. John Pomfret, "China Plans Major Boost in Spending," *The Washington Post*, 2 March 2001.

27. Kristof, "The Rise of China," 72.

28. Avery Goldstein, "Great Expectations: Interpreting China's Arrival," *International Security* 22, 3 (winter 1997/1998): 71.

29. Albert Feuerwerker, *China's Early Industrialization: Sheng Hsuan-Huai and Mandarin Enterprise* (Cambridge, Mass.: Harvard University Press, 1958), and Benjamin Schwartz, *In Search of Wealth and Power: Yen Fu and the West* (Cambridge, Mass.: Harvard University Press, 1964).

Under these conditions China will likely adopt a vision of technological independence that embraces both strategic technology and value-chain strategies. China will maintain and expand strategic plans like the 863 Plan, a technology plan that focuses on "critical" technologies in the civilian economy. At the same time, it will offer market access and intellectual property rights protection to multinational companies willing to transfer technologies that raise China along the production chain. In this view, it does not matter so much what technologies China can master, but how competitive national enterprises are in the global economy. Remaining open to international trade and investment will be essential since Chinese enterprises would have to be part of international production networks. Still, decisions about which technologies to develop and which foreign corporations to allow market access will be based on relative strategic concerns.[30] Adopting this strategy, China will try to maintain autonomy in relation to the states it views as its primary threats to security, most likely the United States and Japan. Short-term economic gains might be sacrificed to maintain technological autonomy from these states. At its broadest, this strategy would reflect a China that views security threats as economic as well as military.

The current leadership under Jiang Zemin appears to be pursuing this strategy, and it allows leaders to maintain several policies simultaneously, attracting the support of both conservative and more reformist factions in the ruling elite. Depending on the international environment, the state could shift between developing strategic industries and core technologies and opening to interaction with the world economy. Elites in both camps could claim to be pursuing the historical mission of using technology to strengthen China. Actually fulfilling this mission and creating a science and technology infrastructure supportive of a modern economy may be more difficult. To succeed in the future, the central government will have to intervene in markets in unaccustomed ways, to support without eliminating individual initiative. In short, the central government will have to act like a good local government.

30. This would be a weak version of "mercantile realism" or the concern for advancing a state's technoeconomic position. Unlike Japan, China is a technology taker, not maker. And Chinese elites have not embraced the idea that the efficacy of military power has decreased in the modern world. But the point about security and relative gains would still hold. Policymakers would fear the use of economic power to limit the sovereignty or independence of China. As a result China would be more sensitive in its trading with the U.S. than it would with France. Eric Heginbotham and Richard Samuels, "Mercantile Realism and Japanese Foreign Policy," *International Security* 22, 4 (spring 1998): 171–203.

Appendix

Location	Number	Date	Unit
Shanghai	S1	11 March 1996	Caohejing High-Technology Development Zone (HTDZ)
	S2	11 March 1996	Caohejing HTDZ
	S3	14 March 1996	Shanghai Academy of Social Science (SASS)
	S4	18 March 1996	Shanghai Science and Technology Commission (STC)
	S5	25 March 1996	Zhangjiang HTDZ
	S6	4 April 1996	Shanghai Xiwang Company
	S7	4 April 1996	Shanghai Xiwang Company
	S8	8 April 1996	District STC
	S9	9 April 1996	Shanghai Patent Office
	S10a	14 April 1996	Fudan University, Applied Math Department
	S10b	14 April 1996	Fudan University, Applied Math Department
	S11	16 April 1996	Software company
	S12	17 April 1996	Shanghai STC
	S13	18 April 1996	Shanghai Users Friend Software
	S13a	18 April 1996	Shanghai Users Friend Software
	S14	18 April 1996	Shanghai Academy of Social Science
	S15	23 April 1996	Huadong Computer Company
	S16	24 April 1996	District STC

	S17	14 June 1996	Fudan University, computer science student
	S18	22 June 1996	Fudan University, computer science student
	S19	24 June 1996	Shanghai Municipal Government
	S20	24 June 1996	SASS
	S21	1 September 1996	District STC
	S22	3 September 1996	Forward Corporation
	S23	10 September 1996	Shanghai S&T University
	S24	16 September 1996	SASS
	S25	17 September 1996	Shanghai Municipal Government
	S26	24 September 1996	Caohejing HTDZ
	S27a	7 October 1996	Jiading HTDZ
	S27b	7 October 1996	Jiading HTDZ
	S27c	7 October 1996	Jiading HTDZ
	S28	14 October 1996	Shanghai Eastern Normal College
	S29	28 October 1996	Shanghai Economic Reform Commission
	S30	3 November 1996	Forward Corporation
	S31	5 November 1996	Shanghai Municipal Government
	S32	12 June 1998	Baoshan Nongovernmental Technology Zone
	S33	16 June 1998	Shanghai Industrial Technology Foundation
	S34	17 June 1998	Shanghai Economic Development Research Institute
	S35	17 June 1998	Forward Corporation
	S36	19 June 1998	Shanghai Industrial Technology Foundation
	S37	23 June 1998	Shanghai High-Technology Zone
	S38	24 June 1998	S&T Development and Exchange Center
	S39	25 June 1998	Software company
Beijing	B1	3 May 1996	Users Friend Software
	B2	5 May 1996	Stone Corporation
	B3	8 May 1996	Founders

B4	8 May 1996	Xiwang Corporation
B5	9 May 1996	University of Science and Technology
B6a	9 May 1996	Chinese Academy of Social Science, Industrial Economics Research Department
B6b	9 May 1996	Chinese Academy of Social Science, Industrial Economics Research Department
B7	10 May 1996	Qinghua University, School of Management
B8	21 January 1997	Qinghua University, School of Management
B9	3 March 1997	Beijing Soft Science Research Center
B10	11 March 1997	National Center for Research in Science and Technology Development
B11	12 March 1997	Beijing Soft Science Research Center
B12	19 March 1997	National Center for Research in Science and Technology Development
B13	26 March 1997	Internet start-up
B14	7 April 1997	Internet start-up
B15	24 April 1997	Beijing Experimental Zone
B16	27 April 1997	State Science and Technology Commission (SSTC)
B17	5 May 1997	Beijing Experimental Zone
B18	8 May 1997	Former employee, software company
B19	12 May 1997	U.S. software company
B20a	15 May 1997	Internet start-up
B20b	15 May 1997	Internet start-up
B20c	15 May 1997	Internet start-up
B21	15 May 1997	Institute of Computing, Chinese Academy of Science (CAS)

	B22	19 May 1997	U.S.-Chinese JV
	B23	21 May 1997	Legend Corporation
	B24	26 May 1997	CAS
	B25	30 May 1997	CAS
	B26	17 June 1997	Science and Technology Daily
	B27	26 June 1997	Legend Corporation
	B28	27 June 1997	Industrial association
	B29	28 June 1997	Consulting company
	B30	3 July 1997	Software company
	B31	7 July 1997	Nongovernmental S&T Entrepreneurs Association
	B32	8 July 1997	Time Corporation
	B33	14 July 1997	JingHai Corporation
	B34	16 July 1997	Stone Corporation
	B35	17 July 1997	Beijing Experimental Zone
	B36	18 July 1997	Beijing Technology News
	B37	24 July 1997	Users Friend Software
	B38	24 July 1997	SSTC
	B39	29 July 1997	Founders
	B40	30 July 1997	Legend Corporation
	B42	5 June 1998	National Center for Research in Science and Technology Development
	B43	6 June 1998	Nongovernmental S&T Entrepreneurs Association
Guangzhou	G1	10 November 1996	Sun Yat Sen University, Biotech Research Center
	G2	10 November 1996	Software company
	G3	11 November 1996	Tianhe High-Technology Zone
	G4	11 November 1996	Suntek
	G5	12 November 1996	Guangzhou STC
	G6	13 November 1996	Guangdong STC
	G7	13 November 1996	Legend/Guangzhou branch office
	G8	14 November 1996	Sun Yat Sen Medical University
	G9	14 November 1996	Internet start-up

G10	29 June 1998	Sun Yat Sen University, Biotech Research Center
G11	30 June 1998	Southern Market Economy Research Institute
G12	30 June 1998	Commercial and Industrial Bureau
G13	3 July 1998	Biotechnology company
G14	4 July 1998	Biotechnology company
G15	6 July 1998	Guangzhou STC
G16	6 July 1998	Suntek
G17	8 July 1998	Tianhe High-Technology Zone
G18	8 July 1998	Suntek
G19	9 July 1998	Internet start-up
G20	9 July 1998	Guangzhou STC
G21	9 July 1998	Sun Yat Sen University, computer science department
Xi'an X1	21 July 1998	Company linked to university
X2	22 July 1998	Xi'an HTDZ
X3	23 July 1998	Software company
X4	23 July 1998	Software company
X5	23 July 1998	Internet start-up
X6	24 July 1998	Xi'an Jiaotong University
X7	24 July 1998	Xi'an North West University
X8	27 July 1998	SeaStar Corporation
X9	28 July 1998	Xi'an HTDZ
X10	28 July 1998	Xi'an Business Incubator
X11	29 July 1998	Future Software
X12	29 July 1998	Taiwanese Joint Venture
X13	30 July 1998	SeaStar Corporation
X14	30 July 1998	Future Software
X15	30 July 1998	Xi'an HTDZ
X16	2 August 1998	Software company
X17	2 August 1998	Xi'an Business Incubator
X18	3 August 1998	Xi'an Jiaotong University

Index

CORNELL STUDIES IN POLITICAL ECONOMY
A series edited by
Peter J. Katzenstein

Pathways from the Periphery: The Politics of Growth in the Newly Industrializing Countries by Stephan Haggard

The Politics of Finance in Developing Countries
edited by Stephan Haggard, Chung H. Lee, and Sylvia Maxfield

Rival Capitalists: International Competitiveness in the United States, Japan, and Western Europe by Jeffrey A. Hart

Reasons of State: Oil Politics and the Capacities of American Government
by G. John Ikenberry

The State and American Foreign Economic Policy
edited by G. John Ikenberry, David A. Lake, and Michael Mastanduno

The Nordic States and European Unity by Christine Ingebritsen

The Paradox of Continental Production: National Investment Policies in North America by Barbara Jenkins

The Government of Money: Monetarism in Germany and the United States
by Peter A. Johnson

Corporatism and Change: Austria, Switzerland, and the Politics of Industry
by Peter J. Katzenstein

Cultural Norms and National Security: Police and Military in Postwar Japan
by Peter J. Katzenstein

Small States in World Markets: Industrial Policy in Europe
by Peter J. Katzenstein

Industry and Politics in West Germany: Toward the Third Republic
edited by Peter J. Katzenstein

Norms in International Relations: The Struggle against Apartheid
by Audie Jeanne Klotz

International Regimes edited by Stephen D. Krasner

Disparaged Success: Labor Politics in Postwar Japan by Ikuo Kume

Business and Banking: Political Change and Economic Integration in Western Europe
by Paulette Kurzer

Power, Protection, and Free Trade: International Sources of U.S. Commercial Strategy, 1887–1939 by David A. Lake

Money Rules: The New Politics of Finance in Britain and Japan
by Henry Laurence

Why Syria Goes to War: Thirty Years of Confrontation by Fred H. Lawson

Remaking the Italian Economy by Richard M. Locke

France after Hegemony: International Change and Financial Reform
by Michael Loriaux

Economic Containment: CoCom and the Politics of East-West Trade
by Michael Mastanduno

Business and the State in Developing Countries
edited by Sylvia Maxfield and Ben Ross Schneider

The Currency of Ideas: Monetary Politics in the European Union
by Kathleen R. McNamara

The Choice for Europe: Social Purpose and State Power from Messina to Maastricht by Andrew Moravcsik

At Home Abroad: Identity and Power in American Foreign Policy
by Henry R. Nau

Collective Action in East Asia: How Ruling Parties Shape Industrial Policy
  by Gregory W. Noble
Mercantile States and the World Oil Cartel, 1900–1939
  by Gregory P. Nowell
Negotiating the World Economy by John S. Odell
Opening Financial Markets: Banking Politics on the Pacific Rim
  by Louis W. Pauly
Who Elected the Bankers? Surveillance and Control in the World Economy
  by Louis W. Pauly
Regime Shift: Comparative Dynamics of the Japanese Political Economy
  by T. J. Pempel
The Politics of the Asian Economic Crisis edited by T. J. Pempel
The Limits of Social Democracy: Investment Politics in Sweden
  by Jonas Pontusson
The Fruits of Fascism: Postwar Prosperity in Historical Perspective
  by Simon Reich
The Business of the Japanese State: Energy Markets in Comparative and Historical Perspective by Richard J. Samuels
"Rich Nation, Strong Army": National Security and the Technological Transformation of Japan by Richard J. Samuels
Crisis and Choice in European Social Democracy
  by Fritz W. Scharpf, translated by Ruth Crowley and Fred Thompson
Winners and Losers: How Sectors Shape the Developmental Prospects of States by D. Michael Shafer
Ideas and Institutions: Developmentalism in Brazil and Argentina
  by Kathryn Sikkink
The Cooperative Edge: The Internal Politics of International Cartels
  by Debora L. Spar
The Hidden Hand of American Hegemony: Petrodollar Recycling and International Markets by David E. Spiro
The Origins of Nonliberal Capitalism: Germany and Japan in Comparison
  edited by Wolfgang Streeck and Kozo Yamamura
Fair Shares: Unions, Pay, and Politics in Sweden and West Germany
  by Peter Swenson
Union of Parts: Labor Politics in Postwar Germany by Kathleen Thelen
Democracy at Work: Changing World Markets and the Future of Labor Unions by Lowell Turner
Fighting for Partnership: Labor and Politics in Unified Germany
  by Lowell Turner
Troubled Industries: Confronting Economic Change in Japan
  by Robert M. Uriu
National Styles of Regulation: Environmental Policy in Great Britain and the United States by David Vogel
Freer Markets, More Rules: Regulatory Reform in Advanced Industrial Countries by Steven K. Vogel
The Political Economy of Policy Coordination: International Adjustment since 1945 by Michael C. Webb

The Myth of the Powerless State by Linda Weiss
The Developmental State edited by Meredith Woo-Cumings
International Cooperation: Building Regimes for Natural Resources and the Environment by Oran R. Young
International Governance: Protecting the Environment in a Stateless Society
by Oran R. Young
Polar Politics: Creating International Environmental Regimes
edited by Oran R. Young and Gail Osherenko
Governing Ideas: Strategies for Innovation in France and Germany
by J. Nicholas Ziegler
Internationalizing China: Domestic Interests and Global Linkages
by David Zweig
Governments, Markets, and Growth: Financial Systems and the Politics of Industrial Change by John Zysman
American Industry in International Competition: Government Policies and Corporate Strategies edited by John Zysman and Laura Tyson